Concrete Masonry Handbook

for Architects, Engineers, Builders

DAVID CHRISTOPHER SHOCK

Concrete Masonry Handbook
for Architects, Engineers, Builders

By Frank A. Randall, Jr., and William C. Panarese

PORTLAND CEMENT  ASSOCIATION

5420 Old Orchard Road, Skokie, Illinois 60077-4321

An organization of cement manufacturers to improve
and extend the uses of portland cement and concrete
through scientific research, engineering field work, and
market development.

Fourth Edition print history

First printing 1976
Second printing 1976
Third printing 1977
Fourth printing 1978
Fifth printing (rev.) 1980
Sixth printing (rev.) 1982
Seventh printing 1985

ISBN 0-89312-001-4

Library of Congress Cataloging in Publication Data

Randall, Frank A., Jr.   1918-
   Concrete masonry handbook for architects, engineers, builders.

   Bibliography: p. 185
   Includes index.
   1. Concrete construction—Handbooks, manuals, etc.
2. Masonry—Handbooks, manuals, etc. I. Panarese,
William C., joint author. II. Title.
TH1461.R36   693.4   75-44908

ACKNOWLEDGMENTS

Many individuals and organizations within the concrete masonry industry have assisted in writing this greatly expanded Fourth Edition of the *Concrete Masonry Handbook.* Trade associations, professional societies, manufacturers, contractors, consultants, and many others have provided research material and illustrations as well as manuscript reviews and commentary. We are particularly grateful for the guidance and fine cooperation of the individuals and/or organizations listed below (alphabetically by organization). This acknowledgment does not necessarily imply approval of the text by any of those listed since the entire handbook was not reviewed by all those listed and since final editorial prerogatives have necessarily been exercised by the Portland Cement Association.

Lucas E. Pfeiffenberger, Besser Company

Ray Cooley, Concrete Masonry Association of California

Frank Erskine, Expanded Shale, Clay and Slate Institute

George A. Miller, International Masonry Institute and Mason Contractors Association of America

Albyn Mackintosh, Mackintosh & Mackintosh, Inc.

James E. Amrhein, Masonry Institute of America

Paul Lenchuk and Henry Toennies, National Concrete Masonry Association

Robert J. Primeau, National Concrete Producers Association (Canada) and Primeau Argo Block Co., Limited

National Research Council of Canada

Thomas A. Holm, Solite Corporation

About the authors. The staff members of the Portland Cement Association responsible for the final manuscript are Frank A. Randall, Jr., and William C. Panarese, structural engineer and manager, respectively, Concrete Technology Section, Engineering Services Department. The following PCA staff (listed alphabetically by name) provided valuable technical assistance, commentary and manuscript reviews:

Melvin S. Abrams, manager, Fire Research Section

Mario J. Catani, director, Building Construction Dept.

Arthur R. Corwin, architect, Building Construction Dept.

Stanley Cumming, manager, Canadian Codes and Standards

Stanley E. Goodwin, structural engineer, Building Construction Dept.

Albert W. Isberner, Jr., senior research engineer, Concrete Materials Research Dept.

Gerald B. Neville, structural engineer, Codes and Standards Dept.

PORTLAND CEMENT ASSOCIATION

PREFACE

Since 1882 when the first concrete block was molded, the concrete masonry industry in the United States and Canada has grown to produce about 4 billion 8×8×16-in. equivalent units each year. Laid end to end at the equator, these units would extend 40 times around the world.

An industry this large and progressive undergoes continuous change and refinement. Since the first edition of this handbook was published in 1951, concrete masonry has come of age as an engineered construction material. Standards, specifications, and codes have been developed in depth. Control joints, joint reinforcement, and manufacturing processes have been developed to minimize wall cracking. Screen wall units and customized architectural masonry have proliferated to create lively, attractive facades. Mortars and grouts are better understood and improved. Knowledge of the acoustical, thermal, and fire-resistant properties of concrete masonry has advanced. Cold-weather construction has been researched and extended. High-strength block and reinforced concrete masonry have been developed to build high-rise load-bearing structures.

These are some of the developments reflected in this comprehensive Fourth Edition, presented to assist architects, engineers, and builders in designing and constructing advanced, economical concrete masonry buildings.

CONTENTS

List of Tables

Chapter 1

GENERAL INFORMATION ON CONCRETE MASONRY UNITS

Concrete masonry units (block and concrete brick) are available in sizes, shapes, colors, textures, and profiles for practically every conceivable need and convenience in masonry construction. In addition, concrete masonry units may be used to create attractive patterns and designs to produce an almost unlimited range in architectural treatments of wall surfaces. The list of current applications is lengthy, but some of the more prominent uses are for:

- Exterior load-bearing walls (below and above grade)
- Interior load-bearing or non-load-bearing walls
- Fire walls, party walls, curtain walls
- Partitions, panel walls, solar screens
- Backing for brick, stone, stucco, and other exterior facings
- Veneer or nonstructural facing for wood, concrete, or masonry
- Fire protection of structural steel members
- Firesafe enclosures of stairwells, elevator shafts, storage vaults, or fire-hazardous work areas
- Piers, pilasters, columns
- Bond beams, lintels, sills
- Floor and roof systems
- Retaining walls, slope protection, ornamental garden walls, and highway sound barriers
- Chimneys and fireplaces (indoor and outdoor)
- Catch basins, manholes, valve vaults
- Paving and turf block

Specifications and Codes

As with every type of building material, there is a multitude of specifications and codes to guide or regulate manufacturers, designers, and builders. A listing of the more prominent references is given at the end of this book.

U.S. and General Requirements

Concrete masonry units are manufactured in the United States to conform to requirements of the American Society for Testing and Materials (ASTM). ASTM specifications classify concrete masonry units according to grade and type. The grade describes the intended use of the concrete masonry units, as shown in Table 1-1, while the type of unit is indicated as either Type I, moisture-controlled, or Type II, non-moisture-controlled. Type I units are specified where drying shrinkage of the block due to loss of moisture could result in excessive stress and cracking of walls.

In common with a number of building materials, concrete shrinks slightly with loss of moisture. The amount of moisture loss is affected by the relative humidity of the surrounding air. After concrete has dried to constant moisture content at one atmospheric condition, a decrease in humidity causes it to lose moisture or an increase causes it to gain. When moist units are placed in a wall and this inherent shrinkage is restrained, as it often is, tensile and shearing stresses are developed that may cause cracks in the walls.

Where concrete block or concrete brick walls will be exposed to low relative humidities, and in areas of the country having exceptionally dry climates, the maximum moisture content permitted at the time of delivery should be lower than when the construction is located in a more humid environment. Furthermore, concrete masonry units have varying inherent drying shrinkage characteristics depending on method of manufacture and materials used. In an attempt to equalize the drying shrinkage, units with low shrinkage characteristics are allowed to have higher moisture

1

Table 1-1. Grades of Concrete Masonry Units for Various Uses *(United States)**

Grade	ASTM C90 or C145 (block)	ASTM C55 (concrete brick)**
N	For general use such as in exterior walls below and above grade that may or may not be exposed to moisture penetration or the weather and for interior walls and backup.	For use as architectural veneer and facing units in exterior walls and for use where high strength and resistance to moisture penetration and severe frost action are desired.
S	Limited to use above grade in exterior walls with weather-protective coatings and in walls not exposed to the weather.	For general use where moderate strength and resistance to frost action and moisture penetration are required.

*Adapted from ASTM C55, Standard Specification for Concrete Building Brick; ASTM C90, Standard Specification for Hollow Load-Bearing Concrete Masonry Units; and ASTM C145, Standard Specification for Solid Load-Bearing Concrete Masonry Units. Not applicable to ASTM C129, Standard Specification for Non-Load-Bearing Concrete Masonry Units.
**Also applicable to solid concrete veneer and facing units larger than brick size, such as split block.

contents than units with high shrinkage. Requirements for moisture-controlled units are shown in Table 1-2.

Compressive strength is an important property of concrete masonry and related to the general usage (described in Table 1-1). Strength requirements are given in Table 1-3.

To understand the significance of the different ASTM strength requirements, it is important to note that net area and shape play an important role. First, the concrete strengths for the respective grades of solid and hollow block or for concrete brick are essentially the same. However, a "solid" block with 75% concrete has more material to resist load than a hollow block with about 50% solid material. Also, a flat concrete brick will have a considerably higher crushing strength than the shell of a hollow block, just as a 2-1/4-in.-high column 4 in. wide will be stronger than an 8-in.-high column 1-1/4 in. wide.

The economic use of concrete masonry over a wide range of applications is made possible by availability of units with a wide range of strengths. Units called "high-strength" block are gaining in popularity, especially in some areas. Their manufacture is slightly different from regular units; they have to be produced more slowly and the selection of block mix must be made more carefully.

High-strength block are not yet defined by a national specification, but they have been considered to fall in the following strength classes:

	Gross area strength, psi	Net area strength (53% solid units), psi
Regular-strength block	1,060	2,000
High-strength block	1,860	3,500
Extra-high-strength block	2,650	5,000

The high-strength block can be manufactured with most aggregates or combinations of aggregates by all block producers. The extra-high-strength block is usually limited to applications where it is required to limit wall thickness in buildings over 10 stories high. Drying shrinkage of high-strength units may be twice that of units made under ASTM C90. On the other hand, absorption decreases with strength and density of the units.*

Density is related to the amount of water absorption of concrete masonry units as shown in Table 1-3. These properties affect construction, insulation, acoustics, appearance, porosity, painting, etc. Absorption affects the quality of mortar needed. If a masonry unit absorbs water too fast, the mortar will need more

_____
*See Ref. 44.

Table 1-2. Moisture Content Requirements for Type I Units (*United States*) *

Linear shrinkage, percent**	Moisture content, percent, based on maximum total absorption (average of 3 units)†		
	Humid	Intermediate	Arid
0.03 or less	45	40	35
From 0.03 to 0.045	40	35	30
0.045 to 0.065 (max.)	35	30	25

*Adapted from ASTM C55, C90, C129, and C145.
**Based upon ASTM Method C426, Standard Method of Test for Drying Shrinkage of Concrete Block, conducted not more than 12 months prior to delivery of units.
†Considering the following humidity conditions at jobsite or point of use: humid, average annual relative humidity above 75%; intermediate, average annual relative humidity 50-75%; and arid, average annual relative humidity less than 50%. [Mean annual relative humidities for the United States are shown in Fig. 6-52, page 141.]

Table 1-3. Strength and Absorption Requirements for Concrete Masonry Units (*United States*)

Type of unit	ASTM designation	Grade of unit	Minimum compressive strength, psi, on average gross area		Maximum water absorption, pcf, based on oven-dry unit weight			
			Average of three units	Individual unit	Lightweight concrete		Medium-weight concrete, 105 to 125 pcf	Normal-weight concrete, 125 pcf or more
					Less than 85 pcf	Less than 105 pcf		
Concrete brick	C55	N	3,500*	3,000*	15	15	13	10
		S	2,500*	2,000*	18	18	15	13
Solid load-bearing block	C145	N	1,800	1,500	—	18	15	13
		S	1,200	1,000	20	—	—	—
Hollow load-bearing block	C90	N	1,000	800	—	18	15	13
		S	700	600	20	—	—	—
Hollow non-load-bearing block	C129	—	350	300	—	—	—	—

*Concrete brick tested flatwise.

water retentivity. This is necessary to give the mason time to place and adjust the block before the mortar stiffens, and to achieve a strong mortar bond. Absorption rate is a specification requirement for clay units but not for concrete masonry. However, the principle applies and control is exercised by limiting the amount of absorption.

Architectural concrete masonry units for interior use should comply with the requirements of ASTM C90 and C145 for hollow and solid load-bearing units, respectively. Architectural concrete masonry units for exterior use are often specified to conform with specifications for concrete building brick, ASTM C55, Type N.

The ASTM specifications for concrete masonry do not fix the weight, color, surface texture, fire resistance, thermal transmission, or acoustical properties of the units.

Canadian Requirements

Concrete masonry units are manufactured in Canada to conform to requirements of the Canadian Standards Association (CSA). CSA Standard A165.1, Concrete Masonry Units, classifies masonry units, except concrete brick, by their physical properties using the four-facet system shown in Table 1-4. All facets are used in order to designate the properties of the unit. For example, H/1000/C/O is a hollow unit with a strength of 1,000 psi (average of 5 units), a density of less than 105 pcf, and an undefined moisture content at the time of shipment.

It is the intention of CSA Standard A165.1 that all types of block may be used indoors. For exterior walls and unprotected foundations, basement or cellar walls the following types are excluded: H/350/N/M, H/350/N/O, H/700/N/M, and H/700/N/O. In addition, a compressive strength of 1,000 psi is considered essential for exterior use.

For concrete brick, the strength, density, absorption, and moisture content classifications are given in Tables 1-5 and 1-6 in accordance with the provisions of CSA Standard A165.2, Concrete Brick Masonry Units.

Fig. 1-1. A modern block plant in operation. In the left background is the concrete mixer. Concrete travels through the large hopper to the block machine. At the right is a curing rack unloading block to a conveyor.

CSA specifications for concrete masonry units do not fix the weight, color, surface texture, fire resistance, thermal transmission, or acoustical properties of the units.

Manufacture

Concrete masonry units are made mainly of portland cement, graded aggregates, and water. Depending upon specific requirements, the concrete mixtures may also contain other suitable ingredients such as an air-entraining agent, coloring pigment, and siliceous and pozzolanic materials.

Mass production has contributed to the relatively low cost of quality concrete masonry units. In many production plants some phases of the manufacturing process are completely automated.

Briefly, the manufacturing process involves the machine-molding of very dry, no-slump concrete into the desired shapes, which are then subjected to an accelerated curing procedure. This is generally followed by a storage or drying phase so the moisture content of the units may be reduced to the specified moisture limits prior to shipment. The concrete mixtures must be carefully proportioned and their consistency controlled so that texture, color, dimensional tolerances, and other desired physical properties are

Table 1-4. Physical Properties of Concrete Masonry Units Except Concrete Brick (*Canada*)*

Facet	Symbol	Property		
First	H S	Solid content		
		Hollow Solid		
Second	350 700 1,000 1,800 2,750 4,000	Minimum compressive strength, psi, calculated on gross area		
		Average of 5 units		Individual unit
		350 700 1,000 1,800 2,750 4,000		300 600 800 1,500 2,300 3,300
Third	A B C N**	Density and water absorption		
		Oven-dry weight of concrete, pcf		Maximum water absorption, pcf
		Over 125 105–125 Less than 105 No limits		10 14 18 No limits
Fourth	M O	Maximum moisture content, percent of total absorption (average of 5 units)		
			Moisture content	
		Linear shrinkage, percent	RH over 75%†	RH under 75%†
		Less than 0.03 0.03 to 0.045 Over 0.045	45 40 35	40 35 30
		No limits where drying shrinkage is not of importance		

*Adapted from CSA A165.1, Concrete Masonry Units. Also note:
1. Classifications H/700/A/O, H/700/B/O, H/700/A/M, and H/700/B/M are not permitted.
2. It is not intended that manufacturers will make masonry units to fit all possible combinations of facets 1, 2, and 3, but rather that the purchaser will be able to select from the manufacturer's range of masonry units a unit which will meet his requirements.
3. Strengths of 2,750 and 4,000 psi are normally produced on request only.
4. When masonry units are used in a dry environment, such as interior partitions, the maximum water absorption limits need not apply.
5. When masonry units are used under humidity conditions considerably lower than climatic humidity, additional precautions against shrinkage may be required.
6. When a particular surface texture, finish, color, uniformity of color, or other special feature is desired, these features should be specified separately by the purchaser.
7. This standard sets out no requirements for fire resistance, thermal transmission, or acoustical properties. The purchaser should specify definite values for any such properties when required.
**Only applicable to classification H350.
†Average annual climatic relative humidity, percent at point of manufacture. [Mean annual relative humidities for Canada are given in Table 6-6, page 141.]

obtained. High-strength units have concrete with higher cement contents and more water, but still have no slump. Automatic machines consolidate and compact these concretes by vibration and pressure, and mold approximately a thousand 8×8×16-in. masonry units (or their equivalent in other sizes) per hour.

Two types of accelerated curing are utilized by the concrete masonry industry, with variations according to local plant requirements and raw materials used. The more common type of curing provides for heating the block in a steam kiln at atmospheric pressure to temperatures ranging from 120 to 180 deg. F. for periods up to 18 hours.* Atmospheric pressure methods may require subsequent accelerated drying treatment or a period of natural drying in the storage yard under protective cover. A variation of this low-pressure curing is the carbonation stage, which is added to reduce the shrinkage characteristics of the masonry units.

The other type of curing is known as autoclaving or high-pressure steam curing.** In this process the units may be subjected to saturated steam at 325 to 375 deg.

*See Ref. 70.
**See Ref. 2.

5

Table 1-5. Strength and Absorption Requirements for Concrete Brick (*Canada*)*

Type**	Minimum compressive strength, psi, on gross area		Maximum water absorption, pcf, based on oven-dry unit weight of concrete		
	Average of 5 units	Individual unit	More than 125 pcf	125 to 105 pcf	Less than 105 pcf
I-3	3,000	2,500			
I-4	4,000	3,500	10	13	15
I-5	5,000	4,500			
II	2,500	2,000	13	15	18

*Adapted from CSA A165.2, Concrete Brick Masonry Units.
**Type I brick is suitable for use in facing masonry exposed to the weather, while Type II brick is intended for use as backup or interior facing masonry and not suitable for exposure to the weather.

Table 1-6. Moisture Content Requirements for Moisture-Controlled Concrete Brick (*Canada*)*

Linear shrinkage, percent**	Moisture content, percent, based on maximum total absorption (average of 5 units)	
	RH over 75†	RH under 75†
Up to 0.03	40	40
0.03 to 0.045	40	35
Over 0.045	35	30

*Adapted from CSA A165.2, Concrete Brick Masonry Units.
**Based upon ASTM Method C426, Test for Drying Shrinkage of Concrete Block.
†Average annual mean relative humidity, percent. [Mean annual relative humidities for Canada are shown in Table 6-6, page 141.]

F. maximum temperatures (80 to 170 psig) in a large cylindrical pressure vessel for various times up to 12 hours. The interval at maximum temperature is preceded by a preset period of 1 to 4 hours, followed by pressure-buildup time of about 3 hours. At the end of the steaming interval it is common practice to drain the condensate and reduce the cylinder pressure as rapidly as possible—within 20 to 30 minutes. By this procedure, stored heat quickly lowers the absorbed moisture content in the units to specification requirements.

For storage or shipment to the building site the units are generally placed in small stacks or "cubes" consisting of layers of fifteen to eighteen 8×8×16-in. units per layer, or the equivalent volume in other sizes. The cubes (40×48×48 in. or 48×48×48 in.) may be assembled on wooden pallets or banded with the bottom layer of block having cores positioned horizontally. Most delivery trucks are equipped with a device for unloading the cubes at storage areas on the building site.

Types

Available to designers are units having a wide selection of weights, sizes, shapes, and exposed surface treatments for virtually any architectural and/or structural function. Manufacturers and their local associations can supply literature that describe available types of units. *In advance of any detailing the designer is urged to determine the sizes, shapes, textures, and other properties in masonry units required for the proposed construction as well as their availability from local producers.*

Normal-Weight and Lightweight Masonry Units

The terms "dense or normal-weight" and "lightweight" are derived from the density of the aggregates used in the manufacturing process. The normal-weight aggregates used are sand and gravel, crushed stone, and air-cooled blast-furnace slag. The lightweight aggregates include expanded shale, clay, and slate; expanded blast-furnace slag; sintered fly ash; coal cinders; and natural lightweight materials such as pumice and scoria. In general, local availability determines the use of any one type of aggregate. In some locations the term "concrete block" has been used to designate only those units made with sand and gravel or crushed stone aggregates. Generally speaking, however, con-

Fig. 1-2. Loading block into autoclave for curing with high-pressure steam.

crete block are made with any of the above aggregates.

The weight class of a concrete masonry unit is based upon the density or oven-dry weight per cubic foot of the concrete it contains. A unit is considered as lightweight if it has a density or unit weight of 105 pcf or less, medium-weight if it has a density between 105 and 125 pcf, and normal-weight if it has a density of more than 125 pcf.

Concretes containing various aggregates range in unit weight approximately as follows:

	Unit weight, pcf
Sand and gravel concrete	130-145
Crushed stone and sand concrete	120-140
Air-cooled slag concrete	100-125
Coal cinder concrete	80-105
Expanded slag concrete	80-105
Expanded clay, shale, slate, and sintered fly ash concretes	75-90
Scoria concrete	75-100
Pumice concrete	60-85
Cellular concrete	25-44

In addition to the concrete density, the weight of an individual concrete masonry unit depends upon the volume of concrete in the unit. The design of the unit in turn affects its volume. Approximate weights of various hollow masonry units may be determined from Fig. 1-3. Use of lightweight aggregates can reduce the weight about 20 to 45%, compared with the weight of

similar units made of normal-weight aggregates, without sacrifice of structural properties. Whether the designer or builder chooses lightweight or normal-weight units generally depends upon availability and the requirements of the structure under consideration.

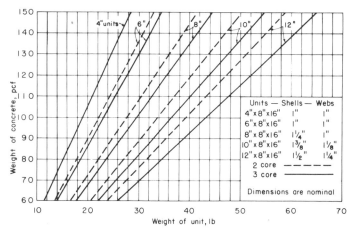

Fig. 1-3. Weights of some hollow units for various concrete densities.

Fig. 1-4. Low-rise residential building of load-bearing concrete masonry.

7

Hollow and Solid Units

Concrete block are classified as hollow or solid units. A hollow unit is defined as one in which the net concrete cross-sectional area parallel to the bearing faces is less than 75% of the gross cross-sectional area. Units having net concrete cross-sectional areas of 75% or more are classified as solid units. The net concrete cross-sectional areas of most concrete masonry units range from 50 to 70% (30 to 50% core area) depending on unit width, face-shell and web thicknesses, and core configuration. Because of their reduced weight and ease of handling, hollow units are in greater use than solid units.

For structural reasons some standards governing concrete masonry units stipulate minimum face-shell and web thicknesses as shown in Table 1-7. ASTM non-load-bearing hollow units are not subject to these requirements, except that the face shell cannot be less than 1/2 in. Furthermore, the face-shell and web thicknesses of CSA Type H/350/N/O units cannot be less than 1/2 in.

Solid masonry units are used primarily for special needs, such as for structures having higher than usual design stresses, for the top or bearing course of load-bearing walls, for increased fire protection, or for catch basin or manhole construction. Concrete brick and some split-block units are made 100% solid (without cores), although some concrete brick designs include a shallow depression, often called a "frog," in one bearing face. However, the net volume of the frog should be not more than 15% of the gross volume of the unit. The purpose of the frog is to reduce weight, provide for better mechanical bond, and prevent the unit from floating when laid in the wall.

Modular Sizes

Concrete masonry unit dimensions are for the most part based on some module, usually 4 or 8 in. From common usage the 3/8-in.-thick mortar joint has become standard. Accordingly, the exterior dimensions of modular units are reduced by the thickness of one mortar joint, 3/8 in. Thus, when laid in masonry the modular units produce wall lengths, heights, and thicknesses that are multiples of the given module. This permits the designer to plan building dimensions and wall openings that will minimize the expense of cutting units on the job.

It is common practice in specifying concrete block to give the wall or block width first, the course (or block) height second, and the block length third, followed by the name of the unit. Dimensions given are nominal, with actual unit dimensions being 3/8 in. less. Fig. 1-5 shows the dimensional details involved in the more popular 8-in. modular grid system.

The nominal block size that dominates the industry is 8×8×16 in. In addition, block is commonly available in widths of 2, 4, 6, 8, 10, and 12 in. Typical nominal block heights are 4 and 8 in., but 12-in. units are also available. Half-length units (nominally 8 in.) and low

Table 1-7. Minimum Thickness of Face Shell and Webs*

Nominal width of unit, in.	Minimum face-shell thickness, in.**	Web thickness	
		Minimum web, in.**	Minimum equivalent web, in./lin.ft.†
3 and 4	3/4	3/4	1-5/8
6	1	1	2-1/4
8	1-1/4	1	2-1/4
10	1-3/8 (1-1/4)††	1-1/8 (1-1/8)	2-1/2 (2-1/2)
12	1-1/2 (1-1/4)††	1-1/8 (1-1/8)	2-1/2 (2-1/2)

*Adapted from ASTM C90 and CSA A165.1.
**Average of measurements on three units taken at the thinnest point.
†Sum of measured thickness of all webs in a unit times 12 divided by length of unit.
††This face-shell thickness is applicable where the allowable design load is reduced in proportion to the reduction in thickness from the basic face-shell thickness shown.

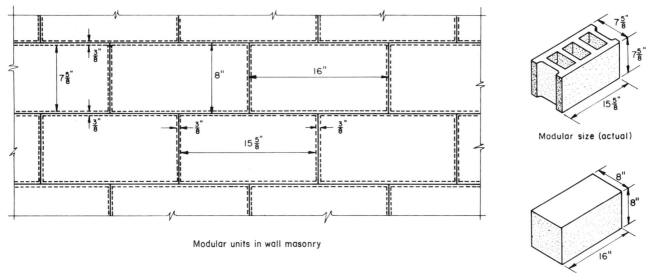

Modular units in wall masonry

Modular size (actual)

Nominal size (usually fictitious)

Fig. 1-5. Dimensional differences between modular- and nominal-sized masonry units.

or half-height units are made as companion items for the mason's convenience in completing various patterns. In some areas the 4-in. module is popular with nominal unit lengths of 8, 12, 16, 20, and 24 in.

Some manufacturers may make units in full nominal dimensions or in stated dimensions other than modular. Whether units are modular or nonmodular, ASTM and CSA standards require that tolerances from the manufacturer's catalogued dimensions not exceed ± 1/8 in. The control in block plants is such, however, that actual variations are seldom greater than ± 1/16 in. The width of split-block units should not be specified within these tolerances. Exposed units should be uniform in size because the neatness of the mortar joint is a big factor in the final appearance and acceptance of an exposed wall.

Core Types

There are many masonry unit design variations. For example, some manufacturers make either two- or three-core units exclusively, while others make some sizes and shapes in both core designs, with the balance of their production in only one or the other core design.

Fig. 1-6 shows some of the variations in core and end designs as applied to 8×8×16-in. concrete masonry units. Some producers make regular stretcher units with flanged ends in the 8-, 10-, and 12-in. widths. Others have adopted, for regular production, single and double plain-ended designs, thereby meeting needs for stretcher, corner, or pier units from single stocks of a given width. In general, all 4-in. and most 6-in.-wide hollow units are made with plain ends but

may contain either two or three cores.

The cores of hollow units are tapered to permit ready stripping in the molding process. Some core designs also include a degree of flaring of the face or web to give a broader base for mortar bedding and for better gripping by the mason. The face shells are sometimes thickened as indicated in Fig. 1-6c; this is done to provide greater tensile strengths at these regions in the finished wall.* End flanges may be grooved or plain. In Canada many producers manufacture only two-core units with grooved mortar recesses on the ends.

Size and Shape Variations

Fig. 1-6 and the remainder of the drawings in this chapter present a sampling of the large number of sizes and shapes of concrete masonry units from which the designer may choose to effect economical and attractive concrete masonry structures. A complete listing is not possible here as some plants produce several hundred different items. Some sizes and shapes are limited to certain areas or are made only on special order. Naturally, there are added costs involved in stocking numerous sizes and shapes, especially if demand is limited. This explains why *there may be limited availability in certain areas.*

Fig. 1-6 shows some types of stretcher and corner units employed in conventional concrete masonry wall construction. Fig. 1-7 includes some of the shapes that are available for the builder's convenience for conven-

---

*See Refs. 3 and 6.

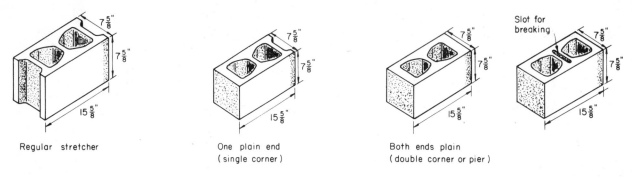

Regular stretcher

One plain end
(single corner)

Both ends plain
(double corner or pier)

(a) Two-core 8x8x16-in. units

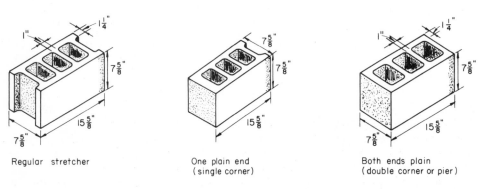

Regular stretcher

One plain end
(single corner)

Both ends plain
(double corner or pier)

(b) Three-core 8x8x16-in. units

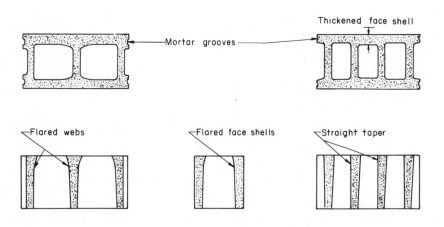

(c) Cross sections

Fig. 1-6. Variations in core and end details of 8×8×16-in. units. Core types shown are used also for 4-, 6-, 10-, and 12-in.-wide units. Generally, 4- and 6-in. units are made with both ends plain. *Consult local producers for specific available units.*

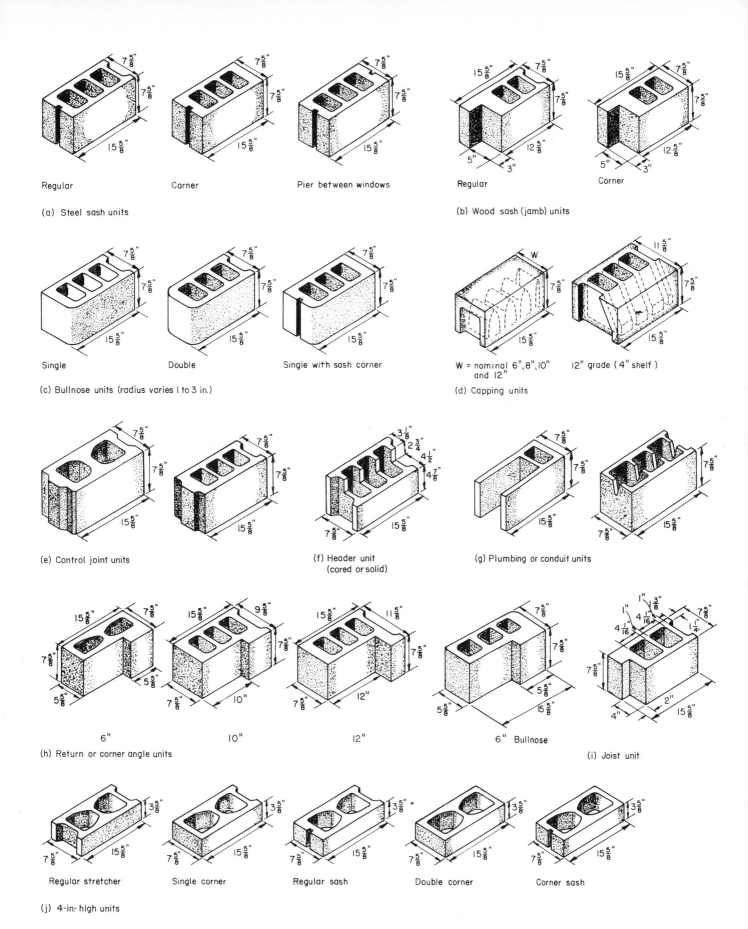

Regular   Corner   Pier between windows

(a) Steel sash units

Regular   Corner

(b) Wood sash (jamb) units

Single   Double   Single with sash corner

(c) Bullnose units (radius varies 1 to 3 in.)

W = nominal 6", 8", 10"   12" grade (4" shelf)
    and 12"

(d) Capping units

(e) Control joint units

(f) Header unit
    (cored or solid)

(g) Plumbing or conduit units

6"   10"   12"   6" Bullnose

(h) Return or corner angle units

(i) Joist unit

Regular stretcher   Single corner   Regular sash   Double corner   Corner sash

(j) 4-in. high units

Fig. 1-7. Some sizes and shapes of concrete masonry units for conventional wall construction. Dimensions shown are actual and used for modular construction with 3/8-in. joints. Half-length units are generally available in shapes shown. Widths range from 4 to 12 in. (nominal) in 2-in. increments for some shapes. *Consult local producers for availability before specifying.*

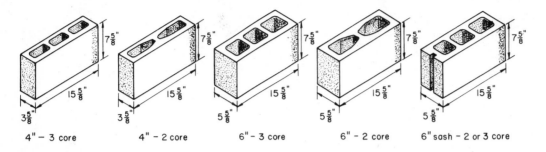

(a) 4-in. and 6-in. partition and backup units.

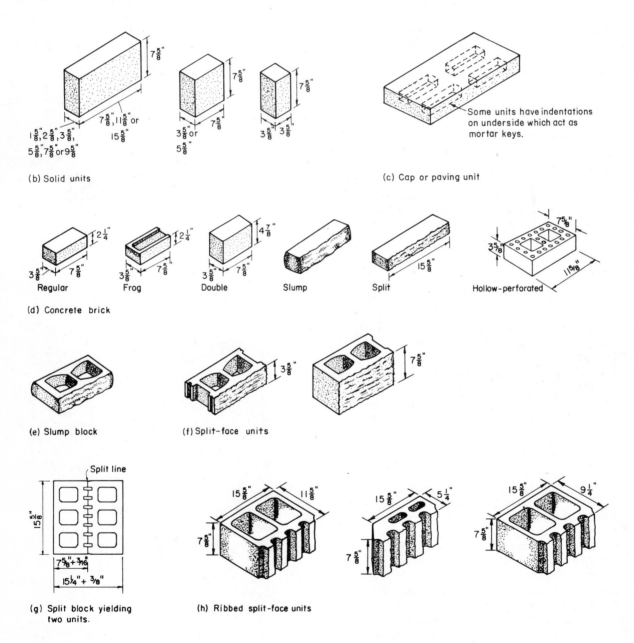

(b) Solid units

(c) Cap or paving unit

Some units have indentations on underside which act as mortar keys.

(d) Concrete brick

Regular   Frog   Double   Slump   Split   Hollow-perforated

(e) Slump block

(f) Split-face units

Split line

(g) Split block yielding two units.

(h) Ribbed split-face units

Fig. 1-8. Some sizes and shapes of concrete masonry units for partition and backup block, solid and cap or paving block, concrete brick, and slump and split units. *Consult local producers for specific available units.*

tional concrete masonry walls or for a particular need. Half-length units are generally available in most of the shapes shown. Alternately, an easily split slotted or kerfed two-core unit may be used (see Fig. 1-6a), or a masonry saw may be used to cut special shapes or shorter lengths from whole units.

Corner blocks have one flush end for use in pilaster, pier, or exposed corner construction. Bullnose blocks have one or more small radius-rounded corners and are used instead of square-edged corner units to minimize chipping. Jamb or sash blocks are used to facilitate the installation of windows or other openings. Capping blocks have solid tops for use as a bearing surface in the finishing course of a wall. Header blocks have a recess to receive the header unit in a composite masonry wall. Return or corner-angle blocks are used in 6-, 10-, and 12-in.-thick walls at corners to maintain horizontal coursing with the appearance of full-length and half-length units.

Fig. 1-8a shows a number of block types available for partition construction and facing unit backup. Solid units and cap or paving units (Fig. 1-8b and c)* are manufactured in a variety of sizes for use as capping units for parapet and garden walls and for use in stepping stones, patios, fireplaces, barbecues, or veneer. They may be used both structurally and nonstructurally. When they are used in reinforced walls, the reinforcing steel is generally placed in grout spaces between wythes.

Some typical concrete brick units are shown in Fig. 1-8d. Concrete brick is sized to be laid with 3/8-in. mortar joints, resulting in modules of 4-in. widths and 8-in. lengths. The thickness of mortar joints is increased slightly so that three courses (three brick and three bed joints) lay up 8 in. high. Some manufacturers produce double brick units.

Slump units (Fig. 1-8d and e) are produced by using a concrete mixture finer and wetter than usual. The concrete brick or block unit is squeezed to give a bulging effect. The rounded or bulging faces resemble handmade adobe, producing a pleasing appearance (Fig. 1-9).

Split brick or block (Fig. 1-8d, f, g, and h) are solid or hollow units that are fractured (split) lengthwise or crosswise by machine to produce a rough stone-like texture (Fig. 1-10). The fractured face or faces, which are exposed when the units are laid, are irregular but sharp, breaking through and exposing the aggregates in the various planes of fracture. By variation of cements, aggregates, color pigments, and unit size, a wide variety of interesting colors, textures, and shapes are produced. The nominal length of split units is 16 in., but half-length units, return corners, and other multiples of 4 in. are obtainable. The solid units are nominally 4 in. wide and available in various modular

*Other units of this type are shown in Chapter 8.

Fig. 1-9. Slump-block garden wall.

Fig. 1-10. Exterior wall of 4-in.-high split-face units.

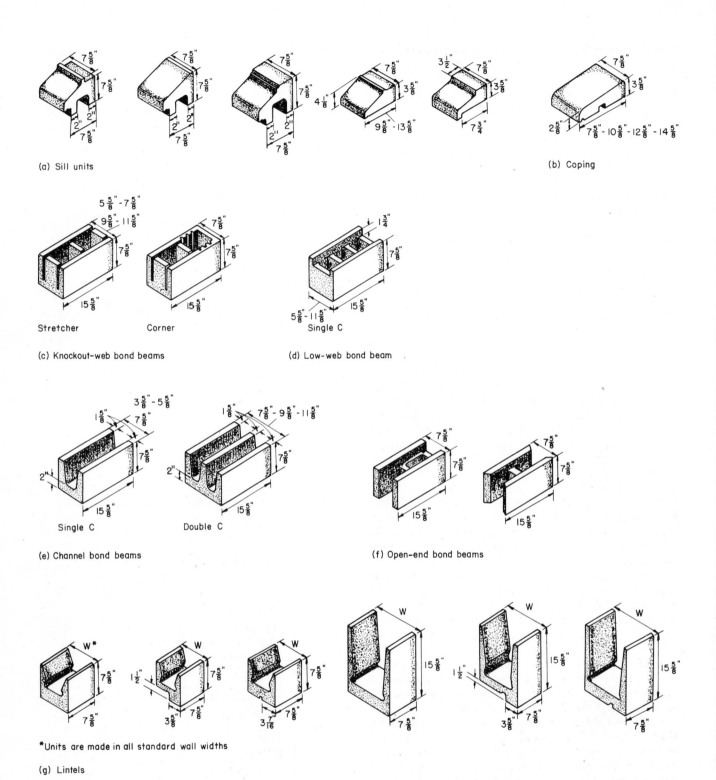

(a) Sill units

(b) Coping

Stretcher   Corner   Single C

(c) Knockout-web bond beams

(d) Low-web bond beam

Single C   Double C

(e) Channel bond beams

(f) Open-end bond beams

*Units are made in all standard wall widths

(g) Lintels

Fig. 1-11. Some sizes and shapes of concrete masonry units for sills, copings, bond beams, and lintels. *Consult local producers for specific available units.*

heights ranging from 1-5/8 to 7-5/8 in. Split solid units are used as a veneering or facing material. Ribbed hollow units, which can be split to produce unusual effects, are useful for through-the-wall applications indoors or out.

Fig. 1-11 shows some of the specialized shapes for constructing window sills, copings, bond beams, and lintels. Reinforced bond beams are useful to minimize cracking due to temperature and moisture change. In areas of recurring earthquakes or hurricanes, where major damage to construction has a high probability of occurrence, reinforced bond beams as well as vertical and horizontal wall reinforcement are mandatory. Reinforced lintels are necessary to bridge over openings for windows and doors. The various sizes and shapes of lintels shown are for different requirements of load capacity, spans, wall widths, and window or door types.

Figs. 1-14 and 1-15 illustrate some types of units made for the construction of pilasters and columns.

Fig. 1-12. Channel bond beam.

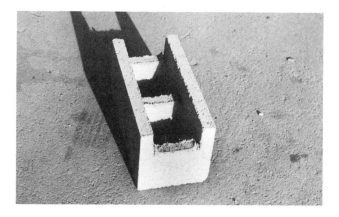

Fig. 1-13. Low-web bond beam.

Fig. 1-16 shows some typical masonry unit shapes made for the construction of sewer manholes, catch basins, valve vaults, and other underground structures. Some unit designs include matching tongue and groove ends. In some areas a similar type of concrete masonry unit may be available for use in the construction of silos and similar containers that must resist internal pressures. These units may be equipped with not only matching tongue and groove ends, but also slots in the bed planes for keyed horizontal joints. Depending upon the service requirements, the whole structure might also be hooped with steel bands.

Screen block or grille units have gained wide popularity as decorative and functional masonry. A few designs are shown in Fig. 1-17. The units available have a wide range of sizes, from 4 to 16 in. square, to meet nearly every need. Though often used mainly for their esthetic value, they also provide excellent balance between privacy and vision from within or without (Fig. 1-18). They diffuse strong sunlight, provide a wind break, and yet permit free flow of air. The decorative value of the units is enhanced by the effects that variations in light and shade produce on their patterns. The principal uses are in decorative building facades, ornamental room dividers and partitions, garden fences, and patio screens. Construction details are given in Chapter 8.

Although there is almost an endless variety of patterns for screen block, *the number available in any locality may be limited.* When a pattern is not available locally, it is often possible for the block plant to rent the mold from the block machine manufacturer or another block producer. The numerous screen block designs available coupled with the possibility of using several orientations of a particular unit in a wall give the designer nearly unlimited opportunity for producing beautiful screen wall effects. Although a wide range of designs with screen block can be obtained, basically the designs include:

1. Units that are a complete pattern in themselves. When laid, the wall forms a panel of small individual repetitive patterns.
2. Units that form part of a pattern. The pattern may require two or sometimes four units to be completed. It is important to consider the esthetic effects of an incomplete pattern if the dimensions of the wall are not a multiple of the dimensions of the pattern.
3. Various types of units that can form an overall pattern in a wall with very interesting and varied effects. Several different patterns are possible

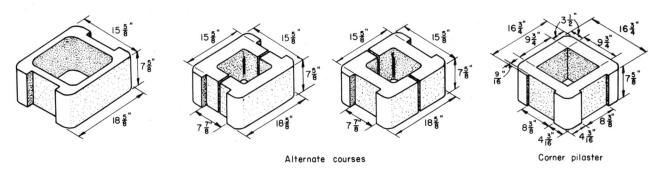

Alternate courses

Corner pilaster

(a) Units for 8-in. walls

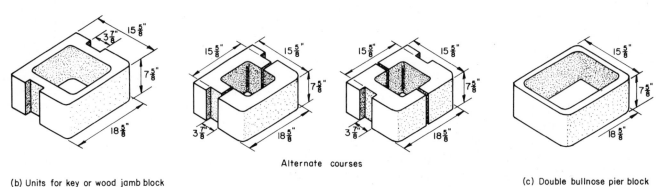

Alternate courses

(b) Units for key or wood jamb block

(c) Double bullnose pier block

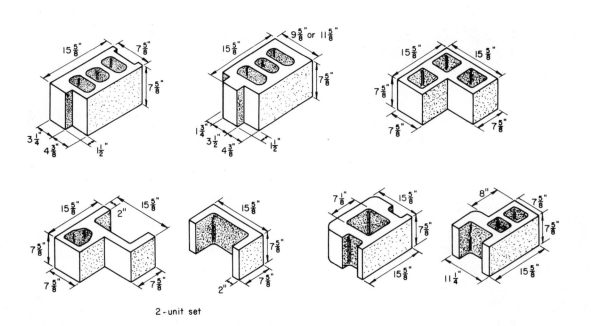

2-unit set

(d) Units for special conditions

Fig. 1-14. Pilaster units. *Consult local producers for specific available units.*

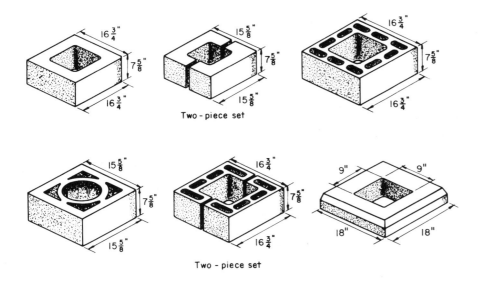

Fig. 1-15. Customized column units. *Consult local producers for specific available units.*

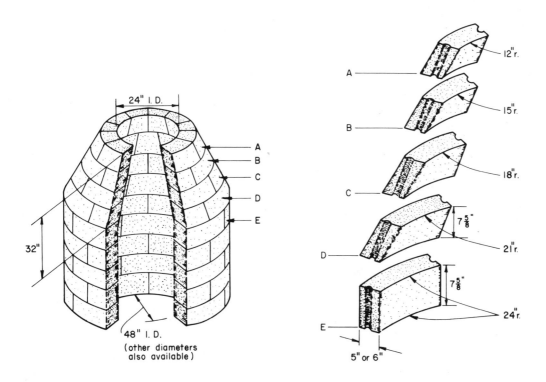

Fig. 1-16. Manhole, catch basin, and valve vault units. *Consult local producers for specific available units.*

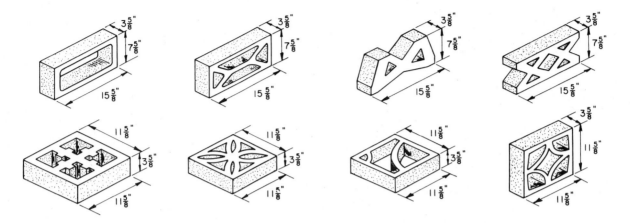

Fig. 1-17. Screen wall units. *Consult local producers for specific available units.*

Fig. 1-18. Screen walls give privacy at the bedrooms in these apartments.

using only two types of these units.

4. Conventional solid or hollow block units, which can be used quite successfully in screen walls if laid with spaces between the units. Hollow block, laid on their ends or sides, also provide an interesting and attractive screen wall.

Architectural concrete masonry units, sometimes known as customized masonry or sculptured units, offer opportunities for almost unlimited architectural freedom. Some of the possibilities are indicated in Fig. 1-19. Patterns and profiles in the block can be achieved with vertical scoring, fluted or ribbed faces, molded angles or curves, projected or recessed faces (Fig. 1-20), or combinations of these surfaces. The designer may select virtually any shape that can be molded vertically within the bounds of the 18×26-in. metal pallet under the block machine. A few block machines have larger pallets. Usually the vertical height of the unit is limited to 7-5/8 in., though some block manufacturers produce units 11-5/8 in. high.

With architectural concrete masonry the selection of the unit profile desired may be closely related to the architectural design of the building. Also, the surface color and texture, whether or not aggregates will be exposed, the type of aggregates to be exposed, whether or not white or buff-colored cement or color pigments will be used, unit dimensions—all of these are selections that the architect may relate to design. Other effects also can be created by using the same block design in different pattern bonds. The play of light and shade on the profiled faces can be varied according to the projected or recessed position in which each block (of the same pattern) is laid. Furthermore, by use of sculptured units alone or in combination with plain units, a variety of geometric patterns in relief can be designed.

Other Sizes, Shapes, and Types

Block scoring refers to a process whereby any block can be given a new apparent face size by saw-cutting the face either horizontally or vertically, or by molding vertical depressions into the face. Scored block are primarily used to achieve a 1/3 to 2/3 face height for ashlar patterns, but it is possible to allow the imagination to run rampant and achieve any number of patterns.

A patented slotted block (Fig. 1-22a) provides unusually high sound energy absorption. The slotted openings molded into the face of the units conduct sound into the cores, acting as damped resonators. Especially effective with sound in the middle and high frequencies, these block are very useful in gymnasiums, factories, bowling alleys, or other places where noise generation is high.

Special lightweight block are made of lightweight insulating concrete in a variety of sizes (Fig. 1-22b).

The weight of an 8×12×24 or 12×8×24-in. unit is comparable to that of a normal-weight block 8×8×16 in.

A special H-shaped block (Fig. 1-22c) is made for reinforced grouted masonry. This unit can be easily placed between tall reinforcing bars.

Other specialized shapes for structural floor and roof systems and paving are discussed in Chapter 8.

Prefacing

Prefaced concrete masonry units offer opportunities for a wide range of colors, patterns, and textures. Prefaced units are sometimes supplied with scored or patterned surfaces—in a variety of colors. Colors may be even or the surface may be dappled. The surface may be extremely smooth or it may be textured. A unit may be prefaced on one or two sides, or sides and end, as required. They also may be produced in a variety of thicknesses, heights, and shapes, with cove bases, bullnose edges, etc. A few examples of prefaced concrete masonry units are shown in Fig. 1-23.

Applicable specifications are: Standard Specification for Prefaced Concrete and Calcium Silicate Masonry Units, ASTM C744, and Prefaced Concrete Masonry Units, CSA Standard A165.3. The more common binders used to face masonry units are resin, resin and inert filler (fine sand), or portland cement and inert filler. Ceramic or porcelainized glazes and mineral glazes are also used, and these units are often called glazed concrete masonry.

There are numerous variations in the resin formulations and in the methods employed in producing resin-base facings. In some facings the resin-aggregate mixture is vibrated into flat pans, applied to faces of hardened block, and cured by application of heat. In others the facing is applied by spraying the face of the hardened block before heat processing.

With the cement-aggregate method, the facing mixture is prepared and vibrated into flat pans. These pans are then inserted into the block machine and the block is cast, with facing and block forming an integral bond. Generally, the heights and lengths of pan-applied facings are 1/8 in. greater than the heights and lengths of the regular modular concrete masonry units to which they are bonded. As a result, the exposed face of the mortar joints must be 1/4 in. thick to preserve modular dimensioning of the masonry wall.

Because the facings are resistant to water penetration, abrasion, and cleaning detergents, prefaced units are used in walls and partitions where decoration, cleanliness, and low maintenance are desirable. This is the case for wainscoting in school corridors and locker rooms, in bottling plants, in special areas of hospitals, and in food processing plants. On the other hand, architects have achieved striking effects by combining

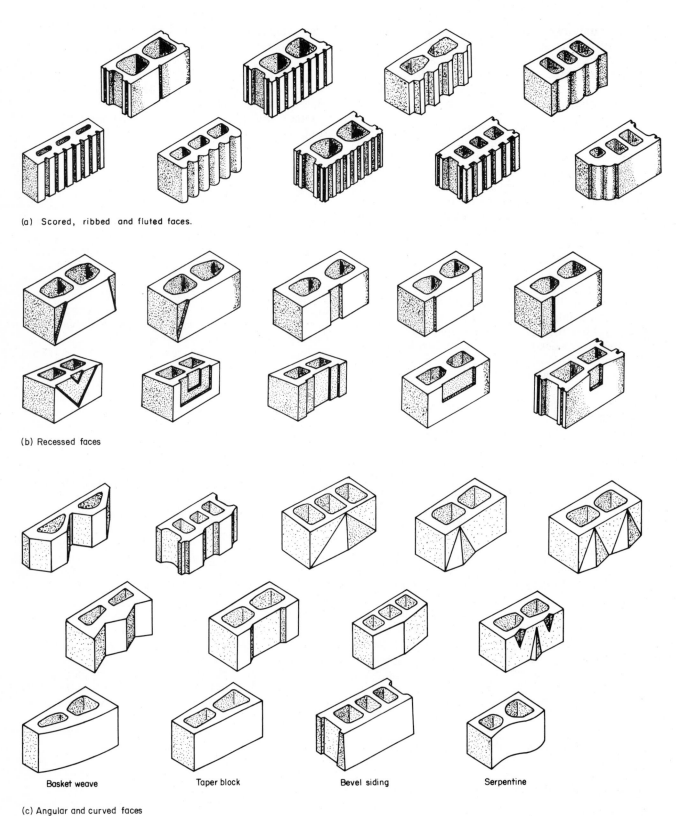

(a) Scored, ribbed and fluted faces.

(b) Recessed faces

Basket weave   Taper block   Bevel siding   Serpentine

(c) Angular and curved faces

Fig. 1-19. Architectural concrete masonry units. *Consult local producers for specific available units.*

Fig. 1-20. Commercial building made with recessed-face block.

Fig. 1-21. Vertical scoring provides a dignified pattern for this office lobby.

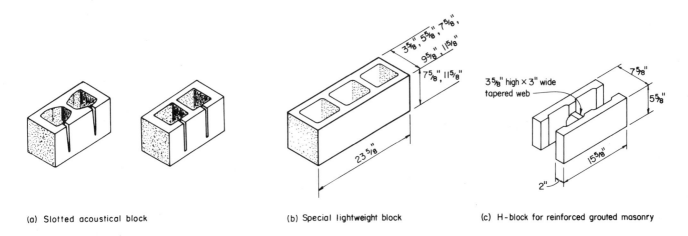

(a) Slotted acoustical block

(b) Special lightweight block

(c) H-block for reinforced grouted masonry

Fig. 1-22. Some special units. *Consult local producers for specific available units.*

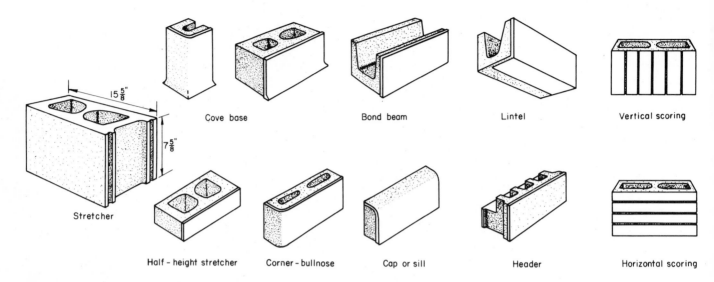

Cove base   Bond beam   Lintel   Vertical scoring

15 5/8"

7 5/8"

Stretcher

Half-height stretcher   Corner-bullnose   Cap or sill   Header   Horizontal scoring

Fig. 1-23. Prefaced units. *Consult local producers for specific available units.*

Fig. 1-24. Glazed concrete masonry units are used at swimming pools where sanitation and a permanently attractive finish are needed.

vividly colored glazed units with regular painted units or with the types of units represented in Fig. 1-19.

In some localities masonry units are available that are prefinished with colored glazes spray-applied in a manner which preserves the original surface texture and sound absorption properties. Thus, they are useful in constructing walls for auditoriums, churches, classrooms, lecture halls, and other areas where pleasing color effects, good acoustical properties, and minimal wall surface maintenance are sought.

Surface Texture

The surface texture of concrete masonry units may be greatly varied to satisfy esthetic requirements or to suit a desired physical requirement. Various degrees of smoothness can be achieved with any aggregate by changes in the aggregate grading, mix proportions, wetness of the mix, and the amount of compaction in molding.

Textures are classified somewhat loosely and with considerable overlapping as open, tight, fine, medium, and coarse. A texture regarded as fine in one locality may be considered medium in another.

An open texture is characterized by numerous closely spaced and relatively large voids between the aggregate particles. Conversely, a tight texture is one in which the spaces between aggregate particles are well filled with cement paste; it has few pores or voids of the size readily penetrated by water and sound.

Fine, medium, and coarse describe the relative smoothness or graining of the texture. A fine texture is not only smooth but made up of small, very closely spaced granular particles. A coarse texture is noticeably large-grained and rough, resulting from the presence of a large proportion of large-size aggregate particles in the surface. Usually, but not necessarily, it will contain substantial-size voids between aggregate particles. A medium texture is one intermediate between fine and coarse; it is an average texture. Examples of several of these textures are shown in Fig. 1-25.

If the concrete masonry surface is to serve as a base for stucco or plaster, a coarse texture is desirable for good bond. Coarse and medium textures provide sound absorption even when painted. The paint, however, must be applied in a manner that does not close all of the surface pores; spray painting is best. A fine texture is preferred for ease of painting.

In regular plant production the texture of concrete masonry units will be fairly uniform from day to day and shipment to shipment, but absolute uniformity is not attainable. The expected range in texture can best be determined from a sample of 10 or more units taken at random from the manufacturer's stockpile.

With a few exceptions manufacturers limit their regular production of concrete masonry units to a single texture for each aggregate type. Otherwise they

Fig. 1-25. Examples of textures of concrete masonry. The top two units were made with lightweight aggregate, the bottom two with normal-weight aggregate.

could not operate economically because of the problems and expense connected with making, stockpiling, and merchandising several classes of units based on small differences in texture. Whether the texture adopted is fine, medium, or coarse will depend to a large extent upon local practice and preference. Some manufacturers will produce other textures on special order provided the order is large enough and the texture desired can be produced with the material and equipment available at a price acceptable to the customer.

A texturing process that may be selected is ground-facing. Ground-face masonry units are produced from normal-weight or lightweight units by grinding off a 1/16- to 1/8-in. layer of concrete from one or both face shells. The process results in a smooth, open-textured surface that shows aggregate particles of varying color to good advantage. Variations in aggregate size, type, and color and the use of integral pigments offer many opportunities for adding interest to the surface in the form of color and texture. Ground-face units in natural or tinted colors are often used in constructing partition walls and corridor walls that are to be exposed without further finishing, except perhaps the application of wax or a colorless sealer. As an alternate to grinding, concrete masonry units may be sandblasted before or after they are placed.

Color

The natural color of concrete masonry varies from light to dark grey to tints of buff, red, or brown, depending upon the color of the aggregate, cement, and other mix ingredients used as well as the method of curing. Color uniformity of masonry units is not controlled unless by special customer order. There may be some variations in the color of any given day's production even though all conditions are apparently alike.

Units also are subject to temporary and permanent changes in color. Colored surfaces are more vivid and darker when wet than when dry. Units made with dark-colored aggregate will slowly become darker with age when subject to weathering because the surface film of cement paste erodes away, exposing the aggregate. While stockpiled at the plant or for long periods at the jobsite, units may undergo slight color changes due to dust and soot lodging in the surface pores. Despite the numerous factors that may affect color, units of a specific type of aggregate and method of curing are generally uniform in color within acceptable limits. This is particularly true when the units for the project or building are from the same manufacturer.

Much concrete masonry construction is painted for architectural or service exposure reasons (see Chapter 7) and so uniform color of the block in these cases is not an important factor. On the other hand, highly

Fig. 1-26. Ground-face units expose the natural color of the aggregate.

pleasing effects have been achieved in building interiors with delicate differences in shades of color of unpainted units (Fig. 1-27). Obviously, such units should be free of stains and other blemishes.

In recent years a noticeable trend has been developing toward the use of colored mortar joints for laying concrete masonry. Many architects prefer mortar joints that are identical to or harmonize with the shade of color of the masonry units involved. Colored mortar is discussed in Chapter 2.

The use of locally available natural sands, cements, and coarse aggregates to produce the desired color is recommended where possible. This will result in a more easily duplicated color in the event of future additions to a structure.

Integral coloration of concrete masonry units is possible through the use of mineral oxide pigments mixed with the concrete before molding. An ever-increasing variety of colored concrete masonry products are being offered. This trend started with pig-

mented standard building block and has since spread to such products as concrete brick, split block, slump block, paving and patio block, screen block, and architectural concrete masonry units.

Standard colors for integrally colored concrete masonry units are tan, buff, red, brown, pink, and yellow. Green can be produced and is quite permanent but expensive, except in light shades. Blue is expensive and not uniform or permanent. Black or grey is another color possibility.

Pigmented concrete masonry units should not be stored in the open. If they are stacked on their sides, greater drying is possible and any soluble lime can combine with carbon dioxide from the air to reduce efflorescence. Efflorescence is a natural foe of coloring. Nonuniformity of coloring can be counteracted by distributing concrete masonry units to random locations on the job, thus intermingling the different shades of the units.

Fig. 1-27. Quietness and beauty are achieved in this church with walls of concrete masonry.

Chapter 2

MORTAR AND GROUT

Mortar for concrete masonry is designed not only to join masonry units into an integral structure with predictable performance properties, but also to: (1) effect tight seals between units against the entry of air and moisture; (2) bond with steel joint reinforcement, metal ties, and anchor bolts, if any, so that they perform integrally with the masonry; (3) provide an architectural quality to exposed masonry structures through color contrasts or shadow lines from various joint-tooling procedures; and (4) compensate for size variations in the units by providing a bed to accommodate tolerances of units.

Grout is an essential element of reinforced concrete masonry. In reinforced load-bearing masonry wall construction, grout is usually placed only in those wall spaces containing steel reinforcement. The grout bonds the masonry units and steel so that they act together to resist imposed loads. In some reinforced load-bearing masonry walls, all cores—with and without reinforcement—are grouted to further increase the wall resistance to loads. Grout is sometimes used in nonreinforced load-bearing masonry wall construction to give added strength. This is accomplished by filling a portion or all of the cores.

Mortar is not grout. Grout can be produced by adding more water to mortar that has a limited amount of lime. Grout and mortar are used differently, have different characteristics, and are handled differently. Thus, the two are not interchangeable.

Mortar

Masonry mortar is composed of one or more cementitious materials, clean well-graded masonry sand, and sufficient water to produce a plastic, workable mixture. By modern specifications the ratios range, by volume, from 1 part of cementitious material to 2-1/4 to 3-1/2 parts of damp, loose mortar sand.

Probably the most important quality of a masonry

mortar is workability because of its influence on other important mortar properties in both the plastic and hardened states. Mortar properties affect the end product—strong, durable, and watertight masonry construction. Since the mortar is an integral part of a concrete masonry structure and some characteristics of mortar materially affect the quality of workmanship obtained, the mortar should be designed and specified with the same care as the masonry unit itself.

Fig. 2-1. Split block in an apartment building. Mortar and the joints have a subtle influence on the architecture.

Desirable Properties of Fresh, Plastic Mortar

Good mortar is necessary for good workmanship and proper structural performance of concrete masonry. Since mortar must bond masonry units into strong, durable, weathertight structures, it must have many desirable properties and the materials must comply with specifications. Desirable properties of mortar while plastic include workability, water retentivity, and a consistent rate of hardening.

Workability

This property of plastic mortar is difficult to define because it is a combination of a number of interdependent, interrelated properties. The interrelated mortar properties considered as having the greatest influence on workability are: consistency, water retentivity, setting time, weight, adhesion, and penetrability.

The experienced mason judges the workability of mortar by the way it adheres to or slides from his trowel. Mortar of good workability should spread easily on the concrete masonry unit, cling to vertical surfaces, extrude readily from joints without dropping or smearing, and permit easy positioning of the unit without subsequent shifting due to its weight or the weight of successive courses. Mortar consistency should change with weather to help in laying the units. A good workable mix should be softer in summer than in winter to compensate for water loss.

Fig. 2-2. Mortar of proper workability is soft but with good body; it spreads readily and extrudes without smearing or dropping away.

Water Retentivity

This is the property of mortar that resists rapid loss of mixing water (prevents loss of plasticity) to the air on a dry day or to an absorptive masonry unit. Rapid loss of water causes the mortar to stiffen quickly, thereby making it practically impossible to obtain good bond and weathertight joints.*

Water retention is an important property and related to workability. A mortar that has good water retentivity remains soft and plastic long enough for the masonry units to be carefully aligned, leveled, plumbed, and adjusted to proper line without danger of breaking the intimate contact or bond between the mortar and the units. When low-absorption units such as split block are in contact with a mortar having too much water retentivity, they may float. Consequently, the water retention of a mortar should be within tolerable limits.

Entrained air, extremely fine aggregate or cementitious materials, or water adds workability or plasticity to the mortar and increases its water retentivity.

Consistent Rate of Hardening

The rate of hardening of mortar due to hydration (chemical reaction) is the speed at which it develops resistance to an applied load. Too rapid hardening may interfere with the use of the mortar by the mason. Very slow hardening may impede the progress of the work since the mortar will flow from the completed masonry. During winter construction, slow hardening may also subject mortar to early damage from frost action. A well-defined, consistent rate of hardening assists the mason in laying the masonry units and in tooling the joints at the same degree of hardness. Uniform joint color of masonry reflects proper hardening and consistent tooling times.

Hardening is sometimes confused with a stiffening caused by rapid loss of water, as in the case of low-water-retention mortars with highly absorptive units. Also, during very hot, dry weather mortar may tend to stiffen more rapidly than usual. In this case the mason may find it advisable to lay shorter mortar beds and fewer units in advance of tooling.

Desirable Properties of Hardened Mortar

Bond

The general term "bond" refers to a specific property that can be subdivided into: (1) the extent of bond, or the degree of contact of the mortar with the concrete masonry units; and (2) the tensile bond strength, or the force required to separate the units. A

---

*See Ref. 26.

chemical and a mechanical bond exist in each category.

Good extent of bond (complete and intimate contact) is important to watertightness and tensile bond strength. Poor bond at the mortar-to-unit interface may lead to moisture penetration through the unbonded areas. Good extent of bond is obtained with a workable and water-retentive mortar, good workmanship, full joints, and concrete masonry units having a medium initial rate of absorption (suction).

Tensile bond strength is perhaps the most important property of hardened mortar. Mortar must develop sufficient bond to withstand the tensile forces brought about by structural, earth, and wind loads; shrinkage of concrete masonry units or mortar; and temperature changes.

Many variables affect bond, including: (1) mortar ingredients, such as type and amount of cementitious materials, water retained, and air content; (2) characteristics of the masonry units, such as surface texture, suction, and moisture content; (3) workmanship, such as pressure applied to mortar bed during placing; and (4) curing conditions, such as temperature, relative humidity, and wind. The effects of some of these variables on bond are discussed below.

All other factors being equal, mortar bond strength is related to mortar composition, especially the cement content. Bond strength of mortar increases as cement content increases.

There is a direct relationship between mortar flow (water content) and tensile bond strength. For all mortars, bond strength increases as water content increases. The optimum bond strength is obtained while using a mortar with the highest water content compatible with workability, even though mortar compressive strength decreases.

Workmanship is paramount in affecting bond strength. The time lapse between the spreading of mortar and the placing of the concrete masonry units should be kept to a minimum because the water content of the mortar will be reduced through suction of the masonry unit on which it is first placed. If too much time elapses before another unit is placed upon it, the bond between the mortar and the unit being placed will be reduced. The mason should not be permitted to realign, tap, or in any way move units after initial placement, leveling, and alignment. Movement breaks the bond between unit and mortar, after which the mortar will not readhere well to the masonry units.

Portland cement in concrete masonry mortars requires a period in the presence of moisture to develop its full strength potential. In order to obtain optimum curing conditions, the mortar mixture should have the maximum amount of mixing water possible with acceptable workability, considering maximum water retention; i.e., lean, oversanded mixtures should be avoided. Freshly laid masonry should be protected

Fig. 2-3. Tensile bond test.

from the sun and drying winds. With severe drying conditions it may be necessary either to wet the exposed mortar joints with a fine water spray daily for about 4 days or to cover the masonry structure with a polyethylene plastic sheet, or to do both.

Durability

The durability of masonry mortar is its ability to endure the exposure conditions. Although aggressive environments and use of unsound materials may contribute to the deterioration of mortar joints, the major destruction is from water entering the concrete masonry and freezing.

In general, damage to mortar joints and to mortar bond by frost action has not been a problem in concrete masonry wall construction above grade. In order for frost damage to occur, the hardened mortar must first be water-saturated or nearly so. After being placed, mortar becomes less than saturated due to the absorption of some of the mixing water by the units. The saturated condition does not readily return except under special conditions, such as: (1) the masonry is constantly in contact with saturated soils; (2) downspouts leak; (3) there are heavy rains; or (4) horizontal ledges are formed. Under these conditions the masonry units and mortar may become saturated and undergo freeze-thaw deterioration.

High-compressive-strength mortars usually have good durability. Because air-entrained mortar will withstand hundreds of freeze-thaw cycles, its use provides good insurance against localized freeze-thaw damage. Mortar joints deteriorated due to freezing and thawing present a maintenance problem generally requiring tuckpointing.

Compressive Strength

The principal factors affecting the compressive strength of concrete masonry structures are the compressive strength of the masonry unit, the proportions of ingredients within the mortar, the design of the structure, the workmanship, and the degree of curing. Although the compressive strength of concrete masonry may be increased with a stronger mortar, the increase is not proportionate to the compressive strength of the mortar.

Tests have shown that concrete masonry wall compressive strengths increase only about 10% when mortar cube compressive strengths increase 130%. Composite wall compressive strengths increase 25% when mortar cube compressive strengths increase 160%.

Compressive strength of mortar is largely dependent on the type and quantity of cementitious material used in preparing the mortar. It increases with an increase in cement content and decreases with an increase in air entrainment, lime content, or water content.

Volume Change

A popular misconception is that mortar shrinkage can be extensive and cause leaky structures. Actually, the maximum shrinkage across a mortar joint is miniscule and therefore not troublesome. This is even truer with the weaker mortars. They have greater creep, i.e. extensibility, and so are better able to accommodate shrinkage.

Appearance

Uniformity of color and shade of the mortar joints greatly affects the overall appearance of a concrete masonry structure. Atmospheric conditions, admixtures, and moisture content of the masonry units are some of the factors affecting the color and shade of mortar joints. Others are uniformity of proportions in the mortar mix, water content, and time of tooling the mortar joints.

Careful measurement of mortar materials and thorough mixing are important to maintain uniformity from batch to batch and from day to day. Control of this uniformity becomes more difficult with the number of ingredients to be combined at the mixer. Pigments, if used, will provide more uniform color if premixed with a stock of cement sufficient for the needs of the whole project. In some areas, colored masonry cements are available.

Tooling of mortar joints at like degrees of setting is important in ensuring a uniform mortar shade in the finished structure. If the joint is tooled when the mortar is relatively hard, a darker shade results than if

Fig. 2-4. Cube compressive strength test.

Fig. 2-5. Ribbed split block wall.

Table 2-1. Proportion Specifications for Mortar*

| Specification | Mortar type | Parts by volume | | |
		Portland cement or portland blast-furnace slag cement	Masonry cement**	Hydrated lime or lime putty
For plain masonry, ASTM C270, CSA A179	M	1	1	—
		1	—	1/4
	S	1/2	1	—
		1	—	Over 1/4 to 1/2
	N	—	1	—
		1	—	Over 1/2 to 1-1/4
	O	—	1	—
		1	—	Over 1-1/4 to 2-1/2
	K	1	—	Over 2-1/2 to 4
For reinforced masonry, ASTM C476†	PM	1	1	—
	PL	1	—	1/4 to 1/2

*Aggregate ratio to be noted: Under the proportion specifications, the total aggregate shall be equal to not less than 2-1/4 and not more than 3 times the sum of the volumes of the cement and lime used. Also note that:
1. Under ASTM C270, Standard Specification for Mortar for Unit Masonry, aggregate is measured in a damp, loose condition and 1 cu.ft. of masonry sand by damp, loose volume is considered equal to 80 lb. of dry sand.
2. Under CSA A179, Mortar for Unit Masonry, aggregate is proportioned on a dry basis and adjusted for bulking.
**Under CSA A179, masonry cement is to be Type H, except that it may be Type L for Type O mortar.
†Standard Specification for Mortar and Grout for Reinforced Masonry.

the joints are tooled when the mortar is relatively soft. Some masons consider mortar joints ready for tooling after the mortar has stiffened but is still thumbprint-hard, with the water sheen gone.

White cement mortar should never be tooled with metal tools because the metal will darken the joint. A glass or plastic joint tool should be used.*

Specifications and Types

The current specifications for mortars for unit masonry are shown in Tables 2-1 and 2-2. Mortar types are identified by proportion or property specification, but not both.

Mortar type classification under the property specification is dependent on the compressive strength of 2-in. cubes using standard laboratory tests (Fig. 2-4) per Standard Specification for Mortar for Unit Masonry (ASTM C270). These laboratory test cubes are

Table 2-2. Property Specifications for Mortar*

| Specification | Mortar type | Compressive strength of 2-in. cubes, psi | |
		At 7 days	At 28 days
For plain masonry, ASTM C270, CSA A179	M	—	2,500
	S	—	1,800
	N	—	750
	O	—	350
	K	—	75
For reinforced masonry, ASTM C476	PM	1,600**	2,500
	PL	1,600**	2,500

*The total aggregate shall be equal to not less than 2-1/4 and not more than 3-1/2 times the sum of the volumes of the cement and lime used. For the property of water retention, allowable initial flow tests range from 100 to 115% (or in the case of ASTM C476 mortars, to a flow of 130% ± 5%) and the flow-after-suction test must exceed 70 or 75%.
**If the mortar fails to meet the 7-day requirement but meets the 28-day requirement, it shall be acceptable.

---

*For a further discussion of tooling as well as the types of mortar joints, see "Tooling Mortar Joints," Chapter 6.

prepared with less water than will be used on the job. Similar cube tests are not intended to be made on the job. Instead, mortar may be tested in the field according to Standard Method for Pre-Construction and Construction Evaluation of Mortars for Plain and Reinforced Masonry (ASTM C780), or after first allowing the units used to absorb water from the mortar.*

The proportion specification identifies mortar type through various combinations of portland cement with masonry cement, masonry cement singly, and combinations of portland cement and lime. The proportion specification governs when ASTM C270 or CSA A179 (Mortar for Unit Masonry) is referred to without noting which specification—proportion or property specification—shall be used.

The ratio of cementitious material to aggregate in the mixture under the property specification may be less than under the proportion specification. This is to encourage preconstruction mortar testing; an economic reward is possible if less cement is required in a mix to meet the strength requirement of the property specification. The testing portion of both specifications is limited to preconstruction evaluation of the mortars.

In both the property and proportion specifications, the amount of water to be used on the job is the maximum that will produce a workable consistency during construction. This is unlike conventional concrete practice where the water-cement ratio must be carefully controlled.

Another physical requirement of both specifications is the water retention limit. In the laboratory it is measured using a "flow-after-suction" test (described in ASTM C91, Standard Specification for Masonry Cement, and CSA A8, Masonry Cement), which simulates the action of absorptive masonry units on the plastic mortar. In performance of a flow test before and after absorptive suction, a truncated cone of mortar is subjected to twenty-five 1/2-in. drops of a laboratory flow table plate (Fig. 2-6). The diameter of the disturbed sample is compared to the original diameter of the conical mortar sample. Allowable initial flow tests range from 100 to 115% (or in the case of ASTM C476 mortars, to a flow of 130% ± 5%) and the flow-after-suction test must exceed 70 or 75%. These values are specified for laboratory test purposes while flow values of 130 to 150% are common for mortar in actual construction.

An interplay of property and proportion specifications is not intended or recognized by specifications for plain masonry, but is mandatory for reinforced masonry under ASTM C476.

Once the design loads, type of structure, and masonry unit have been determined, the mortar type can be selected. No one mortar type will produce a mortar that rates highest in all desirable mortar properties. Adjustments in the mix to improve one property often

Fig. 2-6. Flow test.

are made at the expense of others. For this reason, the properties of each mortar type should be evaluated and the mortar type chosen that will best satisfy the end-use requirements.

In the United States, for plain (nonreinforced) masonry, mortar type is selected by the architect-engineer team on the basis of Table 2-3. For reinforced masonry, the job specification may simply state that the mortar shall meet the requirements of ASTM C476. This allows the mason contractor the option of selecting the individual mixture, PM or PL, that will be used since either type must meet the same strength requirement. This freedom of selection of individual mixtures by the mason contractor is favored since workability of mixtures and availability of cementitious materials vary with geographical areas.

In Canada, the selection of mortars for plain and reinforced masonry depends on whether or not the masonry is based on engineering analysis of the structural effects of the loads and forces acting on the structure. For masonry that is based on engineering analysis, Type M, S, or N mortar is permitted. For masonry *not* based on engineering analysis, Type M, S, N, O, or K is permitted, except that Types O and K are not allowed where the masonry is to be: (1) directly in contact with soil, such as in a foundation wall, or (2) exposed to the weather on all sides, such as in a parapet wall, balustrade, chimney, and steps and landings.

From the preceding paragraphs it can be seen that a choice of mortar type still exists. Where an engineering analysis is required, the type of mortar selected will, in conjunction with the compressive strength of

---

*See "Testing Field Mortar," Chapter 6.

the concrete masonry units, determine the allowable stresses for the wall. It is not always necessary to use a Type M mortar for high strength because many building codes rate Type S in the wall as giving equally high strength. Moreover, Type S or N mortar has more workability, water retention, and extensibility.

The choice of using masonry cement or a portland cement and lime combination is largely a matter of economics and convenience. Either will produce mortar with acceptable properties as long as the specifications are met. Masonry cements provide all cementitious materials required for a masonry mortar in one bag. The quality and appearance of mortars made from masonry cement are consistent because the masonry cement materials are mixed and ground together before being packaged. Consequently, masonry cement mortars are less subject to variations from batch to batch than mortars produced from combining ingredients on the job.

Colors

Pleasing architectural effects with color contrast or harmony between masonry units and joints are obtained through the use of white or colored mortars.

White mortar is made with white masonry cement, or with white portland cement and lime, and white sand. For colored mortars, the use of white masonry cement or white portland cement instead of the normal grey cements not only produces cleaner, brighter colors but is essential for making pastel colors such as buff, cream, ivory, pink, and rose.

Integrally colored mortar may be obtained through use of color pigments, colored masonry cements, or colored sand. Brilliant or intense colors are not generally attainable in masonry mortars. The color of the mortar joints will depend not only on the color pigment, but also on the cementitious materials, aggregate, and water-cement ratio.

Table 2-3. Guide to the Selection of Mortar Type (United States)*

Kind of masonry	Types of mortar
Foundations:	
Footings.	M, S
Walls of solid units	M, S, N
Walls of hollow units	M, S
Hollow walls.	M, S
Masonry other than foundation masonry:	
Piers of solid masonry.	M, S, N
Piers of hollow units	M, S
Walls of solid masonry.	M, S, N, O
Walls of solid masonry, other than parapet walls, not less than 12 in. thick or more than 35 ft. in height, supported laterally at intervals not exceeding 12 times the wall thickness.	M, S, N, O, K
Walls of hollow units; load-bearing or exterior, and hollow walls 12 in. or more in thickness	M, S, N
Hollow walls, less than 12 in. thick where assumed design wind pressure:	
1. exceeds 20 psf.	M, S
2. does not exceed 20 psf	M, S, N
Linings of existing masonry, either above or below grade.	M, S
Masonry other than above	M, S, N

*Adapted from *American Standard Building Code Requirements for Masonry,* National Bureau of Standards (Ref. 57). Mortar types changed to ASTM C270 designations. This table also appears in Ref. 26.

Pigments must be thoroughly dispersed throughout the mix. To determine if mixing is adequate, some of the mix is flattened under a trowel. If streaks of color are present, additional mixing is required. For best results, the pigment should be premixed with the cement in large controlled quantities. Colored masonry cement produced by cement plants is available in many areas.

As a rule, color pigments should be of mineral oxide composition and contain no dispersants that will slow or stop the portland cement hydration. Iron, manganese, chromium, and cobalt oxides have been used successfully. Zinc and lead oxides should be avoided because they may react with the cement. Carbon black may be used as a coloring agent to obtain dark grey or almost black mortar, but lampblack should not be used. Carbon black should be limited to 3% by weight of the portland cement; durability of this mortar may be deficient.

It is recommended to use only those pigments that have been found acceptable by testing and experience. The following is a guide to the selection of coloring materials:

Red, yellow, brown, black . . Iron oxide
Green . . . . Chromium oxide
Blue . . . . Cobalt oxide
Black or grey . . . . Carbon pigments

Only the minimum quantity of pigment that will produce the desired shade should be used. An excess of pigment, more than 10% of the portland cement by weight, may be detrimental to the strength and durability of the mortar. The quantity of water used in mixing colored mortar should be accurately controlled. The more water, the lighter the color. Retempering or the addition of water while using colored mortar should be avoided. Thus, any mortar not used while plastic and workable should be discarded.

Variations in the color of the materials are such as to make a color formula only approximate. Best results are obtained by experiment. Test panels should be made using the same materials and proportions as intended for use in the actual work, and the panels stored under conditions similar to those at the jobsite for about 5 days. Panels will have a darker shade when wet than when dry.

Fading of colored mortar joints may be caused by efflorescence,* the formation of a white film on the surface (Fig. 2-7). Efflorescence is more visible on a colored surface. The white deposits are caused by soluble salts that have emerged from below the surface, or by calcium hydroxide liberated during the setting of the cement and then combining with atmospheric carbon dioxide forming carbonate compounds. Good color pigments do not effloresce or contribute to efflorescence. Efflorescence may be removed with a light sandblasting or stiff-bristle brush.

_____

*See Ref. 43.

Fig. 2-7. A severe case of efflorescence.

Table 2-4. Aggregate Gradation for Masonry Mortar

Sieve size No.	Gradation specified, percent passing		
	ASTM C144*		CSA A82.56**
	Natural sand	Manufactured sand	
4	100	100	100
8	95 to 100	95 to 100	95 to 100
16	70 to 100	70 to 100	60 to 100
30	40 to 75	40 to 75	35 to 80
50	10 to 35	20 to 40	15 to 50
100	2 to 15	10 to 25	2 to 15
200	—	0 to 10	—

*Additional requirements: Not more than 50% shall be retained between any two sieve sizes nor more than 25% between No. 50 and No. 100 sieve sizes. In those cases where an aggregate fails to meet the gradation limit specified, it may be used if the masonry mortar will comply with the property specification of ASTM C270 (Table 2-2).

**Fine aggregate shall be so graded that neither the proportion of particles finer than a No. 16 sieve and coarser than a No. 30 sieve nor the proportion of particles finer than a No. 30 sieve and coarser than a No. 50 sieve exceeds 50%.

Components

Cementitious Materials

Foremost among the factors that contribute to good mortar is the quality of the mortar ingredients. The following material specifications of the American Society for Testing and Materials (ASTM) or the Canadian Standards Association (CSA) are applicable:

- Masonry cement—ASTM C91, CSA A8 (Type H or L)
- Portland cement—ASTM C150 (Type I, IA, II, IIA, III, or IIIA), CSA A5 (Normal, Moderate, or High-Early Strength)
- Portland blast-furnace slag cement—ASTM C595 (Type IS or IS-A)
- Portland pozzolan cement*—ASTM C595 (Type IP or IP-A when fly ash is the pozzolanic material)
- Hydrated lime for masonry purposes—ASTM C207 (Type S or N), CSA A82.43
- Quicklime for structural purposes—ASTM C5, CSA A82.42

Masonry Sand

The quantity of sand required to make 1 cu.ft. of mortar may be as much as 0.99 cu.ft.; hence, the sand has considerable influence on the mortar properties. Masonry sand for mortar should comply with the requirements of ASTM C144 (Standard Specification for Aggregate for Masonry Mortar) for masonry construction within the United States and CSA A82.56 (Aggregate for Masonry Mortar) within Canada. These specifications include both natural and manufactured sands. Sand should be clean, well-graded, and meet the gradation requirements listed in Table 2-4.

Sands with less than 5 to 15% passing the Nos. 50 and 100 sieves generally produce harsh or coarse mortars possessing poor workability; they also result in mortar joints with low resistance to moisture penetration. On the other hand, sands finer than those permitted by the above specifications yield mortars with excellent workability, but they are weak and porous.

For mortar joints that are less than the conventional 3/8-in. thickness, 100% of the sand should pass the No. 8 sieve and 95% the No. 16 sieve. For joints thicker than 3/8 in., the mortar sand selected should have a fineness modulus** approaching 2.5 or a gradation within the limits of concrete sands (fine aggregate) shown in ASTM C33, Standard Specification for Concrete Aggregates, or CSA A23.1, Concrete Materials and Methods of Concrete Construction.

All cementitious materials and aggregates should be stored in such a manner as to prevent wetting, deterioration, or intrusion of foreign material. Brands of cementitious materials and the source of supply of sand should remain the same throughout the entire job.

---

*For use with mortar property specification only.

**Fineness modulus equals the sum of the cumulative percentages *retained* on the standard sieves, divided by 100. The higher the fineness modulus, the coarser the sand.

Water

Water intended for use in mixing mortar should be clean and free of deleterious amounts of acids, alkalies, and organic materials. Some potable waters contain appreciable amounts of soluble salts such as sodium and potassium sulfate. These salts can contribute to efflorescence later. Also, a water containing sugar would retard the set. Thus, the water should be fit to drink but investigated if it contains alkalies, sulfates, or sugars.

Admixtures

Although water-reducers, accelerators, retarders, and other admixtures are used in concrete construction, their use in masonry mortar may produce adverse effects on the normal chemical reaction between cement and water, especially during the early periods after mixing and when the water is available for hydrating the portland cement. Whenever admixtures are considered for use in masonry, it is recommended that the admixture be laboratory-tested in the construction mortars at the temperature extremes requiring their use. The construction site should be inspected to ensure their satisfactory performance under the conditions that prevail.

Materials that retard the hydration process are particularly undesirable because they reduce strength development and increase the potential toward efflorescence. Air-entraining admixtures and calcium chloride accelerators are discussed later (Chapter 5) as admixtures for cold-weather masonry construction.

Measuring Mortar Materials

Measurement of masonry mortar ingredients should be completed in a manner that will ensure the uniformity of mix proportions, yields, workability, and mortar color from batch to batch. Aggregate proportions are generally expressed in terms of loose volume, but experience has shown that the amount of sand can vary due to moisture bulking.

Fig. 2-8 shows how loose sand with varying amounts of surface moisture occupies differing volumes. Fig. 2-9 has the same data in another form for the fine and coarse sands, and shows the density of the sand. Loose, damp sand may consist of from 76 to 105 pcf of sand itself, plus the weight of the water. CSA A179 requires that the volume of sand be adjusted to the amount of bulking. ASTM C270 merely states that a cubic foot of loose, damp sand contains 80 lb. of dry sand.

Ordinary sands will absorb water amounting to 0.4 to 2.3% of the weight of the sand. In the field, damp sands usually have 4 to 8% moisture, and so most of the water is on the surface of the sand.

Aside from the sand, other mortar ingredients are

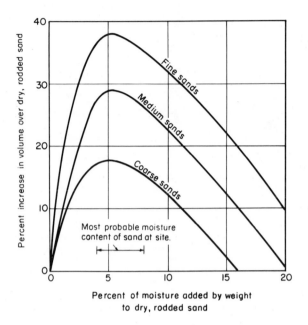

Fig. 2-8. Volume of loose, damp sand.

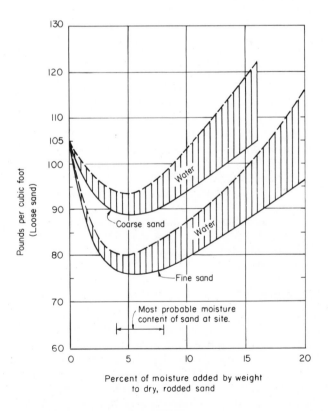

Fig. 2-9. Weight of loose, damp sand.

often sold in bags labeled only by weight. Since mortar is proportioned by volume, it is necessary to know:

	Unit weight, pcf
Portland cement............	94
Masonry cement	70
Hydrated lime (dry)	40
Hydrated lime (putty)........	80

The usual practice of merely measuring sand by the shovel can result in excessive oversanding or undersanding of the mix. For a more positive control, the following method is suggested: Construct one or two wooden boxes 12 in. square and 6 in. deep and use them to measure the sand required in a batch. Add the cement or lime by the bag. Then add water, measuring by the pail. When the desired consistency of mix is determined, mark the level of the mortar in the mixing drum. Use that as the mark for later batches when sand will be added by the shovelful. Repeat the measuring process halfway through the day or whenever the inspector requests it.

A recent innovation is the process of dry-batching all mortar ingredients. This overcomes the problem of adjusting the mix for moisture content of the sand and the usual problem of consistent portions of sand and cementitious materials. In dry-batching, the cementitious materials and dried sand are accurately weighed and mixed at a central plant before delivery in a sealed truck to the site, where the mixture is conveyed into a sealed weathertight hopper. When the mason contractor is ready for mortar, he has only to add the water and mix. It is apparent that this method offers great convenience and can result in very uniform mortar mixes.

Mixing of Mortar

To obtain good workability and other desirable properties of plastic masonry mortar, the ingredients must be thoroughly mixed.

Mixing by Machine

Except possibly on very small jobs, mortar should be machine-mixed. A typical mortar mixer has a capacity of 4 to 7 cu.ft. Conventional mortar mixers are of rotating spiral or paddle-blade design with tilting drum. After all batched materials are together, they should be mixed from 3 to 5 minutes. Less mixing time may result in nonuniformity, poor workability, low water retention, and less than optimum air content. Longer mixing times may adversely affect the air contents of those mortars containing air-entraining cements, particularly during cool or cold weather. Longer mixing times may also reduce the strength of the mortar.

Fig. 2-10. Mortar can be dry-batched at a central plant, delivered in a sealed truck, and stored at the jobsite in a weathertight hopper.

Fig. 2-11. For best results, mortar should be mixed with a power mortar mixer.

Batching procedures will vary with individual preferences. Experience has shown that good results can be obtained when about three-fourths of the required water, one-half of the sand, and all of the cementitious materials are briefly mixed together. The balance of the sand is then charged and the remaining water added. The amount of water added should be the maximum that can be tolerated and still attain satisfactory workability. Mixing is carried out most effectively when the mixer is charged to its design capacity. Overloading can impair mixing efficiency and mortar

uniformity. The mixer drum should be completely empty before charging the next batch.

Mixing by Hand

When hand-mixing of mortar becomes necessary, such as on small jobs, all the dry materials should first be mixed together by hoe, working from one end of a mortar box (or wheelbarrow) and then from the other. Next, two-thirds to three-fourths of the required water is mixed in with the hoe and the mixing continued as above until the batch is uniformly wet. Additional water is carefully added with continued mixing until the desired workability is attained. The batch should be allowed to stand for approximately 5 minutes and then be thoroughly remixed with the hoe.

Retempering

Fresh mortar should be prepared at the rate it is used so that its workability will remain about the same throughout the day. Mortar that has been mixed but not used immediately tends to dry out and stiffen. However, loss of water by absorption and evaporation on a dry day can be reduced by wetting the mortarboard and covering the mortar in the mortar box, wheelbarrow, or tub.

If necessary to restore workability, mortar may be retempered by adding water; thorough remixing is then necessary. Although small additions of water may slightly reduce the compressive strength of the mortar, the end effect is acceptable. Masonry built using plastic mortar has a better bond strength than masonry built using dry, stiff mortar.

Mortar that has stiffened because of hydration hardening should be discarded. Since it is difficult to determine by sight or feel whether mortar stiffening is due to evaporation or hydration, the most practical method of determining the suitability of mortar is on the basis of time elapsed after mixing. Mortar should be used within 2-1/2 hours after mixing.

If colored mortar is used, no retempering should be permitted. Additional water may cause a significant lightening of the mortar.

Mixing During Cold Weather

When masonry construction is carried on during periods of freezing weather, facilities should be available and ready for preparing the mortar and protecting fresh masonry work against frost damage. The preparation of mortar for use under these conditions is particularly important. Chapter 5 goes into more detail on these matters.

Grout

Masonry grout is composed of a mixture of cementitious material and aggregate to which sufficient water

Fig. 2-12. To restore workability, mortar may be retempered.

is added to cause the mixture to flow readily into the masonry cores and cavities without segregation.

Selection

The fineness or coarseness of a grout is selected on the basis of the size of the grout space as well as the height of the lift to be grouted. Building codes and standards differ on specific values of maximum grout aggregate size versus clear opening, and so the governing document should be consulted.

For fine grout (grout without coarse aggregate), generally the smallest space to be grouted should be at least 2×3 in. or 3/4 in. wide, as occurs in the collar joint of two-wythe wall construction (the joint between inner and outer wythes) in low-lift* grouting.

In high-lift* grouting where the smallest horizontal dimension is 3 in., a 1/2-in. maximum-size coarse aggregate (or pea gravel) may be used in the grout. Some specifying agencies stipulate that 3/4-in. maximum-size coarse aggregate may be used when the grout space is 4 in. or greater. The maximum size of the aggregate and consistency of the mix should be selected considering the particular job conditions to ensure satisfactory placement of the grout fill and proper embedment of the reinforcement.

Specifications

Grout for use in concrete masonry walls should comply with the requirements of Standard Specifica-

tion for Mortar and Grout for Reinforced Masonry, ASTM C476, summarized in Table 2-5. The National Building Code of Canada has almost identical provisions.

Aggregates should meet the requirements of ASTM C404, Standard Specification for Aggregates for Masonry Grout, or ASTM C144, Standard Specification for Aggregate for Masonry Mortar.

All of the materials included in ASTM C476 are satisfactory for use in grout. Most projects using large volumes of grout obtain the grout from a ready mixed concrete producer; the use of lime then becomes uneconomic because of the expense in handling.

Admixtures

Practice has shown that a "grouting aid" admixture may be desirable when the concrete masonry units are highly absorbent. The desired effect of the grouting aid is to reduce early water loss to the masonry units, to promote bonding of the grout to all interior surfaces of the units, and to produce a slight expansion sufficient to help ensure complete filling of the cavities.

The use of calcium chloride is strongly discouraged in grout because of possible excessive corrosion of reinforcement, metal ties, or anchors.

----

*See sections under "Building Reinforced Walls," Chapter 6.

Table 2-5. Grout Proportions*

Type	Parts by volume			
	Portland cement, portland blast-furnace slag cement, or portland-pozzolan cement	Hydrated lime or lime putty	Aggregate measured in a damp, loose condition	
			Fine	Coarse
Fine grout	1	0 to 1/10	2-1/4 to 3 times the sum of the volumes of the cementitious materials	—
Coarse grout	1	0 to 1/10	2-1/4 to 3 times the sum of the volumes of the cementitious materials	1 to 2 times the sum of the volumes of the cementitious materials

*Adapted from ASTM C476. Note that the National Bureau of Standards' *Building Code Requirements for Reinforced Masonry* (Ref. 56) allows 1/4 part of lime by volume. It also specifies that the sum of the volumes of the fine and coarse aggregate should not exceed 4 times the sum of the separate volumes of the cement and lime used.

Strength

The mix proportions in Table 2-5 will, in most cases, produce grouts with a compressive strength of 600 to 2,500 psi at 28 days, depending on the amount of mixing water used, when sampled and tested by the conventional laboratory methods used for sampling and testing mortar and concrete. However, the actual in-place compressive strength of the grout generally will exceed 2,500 psi because, under ordinary conditions, some of the mixing water contained in the grout will be absorbed by the concrete masonry during the time the grout is being placed and prior to setting and hardening. This absorption of moisture, in effect, reduces the water-cement ratio of the in-place grout and increases the compressive strength. The moisture, which is absorbed and held by the surrounding concrete masonry during the period immediately following placement of the grout, helps to maintain the grout in the moist condition needed for satisfactory cement hydration and strength gain.

Some building codes specify a minimum compressive strength of 2,000 psi for grout at 28 days when tested according to the method outlined later (under "Sampling and Testing").

Slump

All grout should be of fluid consistency but only fluid enough to pour or pump without segregation. It should flow around the reinforcing bars and into all joints of the concrete masonry, leaving no voids. There should be no bridging or honeycombing of the grout.

The consistency of the grout as measured using a slump test should be based on the rate of absorption of the concrete masonry units and temperature and humidity conditions. Slump is not specified in codes, standards, or specifications. However, when the slump is measured using ASTM C143, Standard Method of Test for Slump of Portland Cement Concrete, the desired slump is 8 in. for units with low absorption and up to 10 in. for units with high absorption.

Mixing

Wherever possible, grout should be batched, mixed, and delivered in accordance with the requirements for ready mixed concrete, e.g. ASTM C94, Standard Specification for Ready-Mixed Concrete. Because of its high slump, ready mixed grout should be continuously agitated after mixing until placement.

Mixing of grout on the jobsite is usually not recommended unless unusual conditions exist that would require special consideration. When a batch mixer is used on the jobsite, all materials should be thoroughly mixed for a minimum of 5 minutes. Grout not placed within 1-1/2 hours after water is first added to the batch should be discarded.

Curing

The high water content of the grout and the partial absorption of this water by concrete masonry will generally provide adequate moisture within the masonry for curing both the mortar and grout. In dry areas where the masonry is subjected to high winds, some moist-curing (such as wetting the structure) may be necessary.

Grouts placed during cold weather are particularly vulnerable to freezing during the early period after grouting because of their high water content. Chapter 5 discusses methods of achieving good curing during cold weather.

Sampling and Testing

Field-testing to establish quality control of the grout as used at the jobsite has merit. Some building codes require that compressive strength tests of grout be made, while others leave it to the option of the specifier.

The number of grout samples to be taken should be specified prior to the start of construction. If not specified, four prism specimens should be cast for each 30 cu.yd. of grout or fraction thereof being placed each day. Also, a sample should be taken whenever there is any change in mix proportions, method of mixing, or materials used.

Grout specimens for compressive tests should not be cast in the usual cylinder molds used for concrete samples because the high water content of the grout would cause low strength results not indicative of the actual in-the-wall strength. Instead, they should be cast in molds formed with concrete masonry units having the same absorption characteristics and moisture content as the units being used in the construction. This simulates in-the-structure conditions, where water from the grout is freely absorbed by the units, thus reducing the water-cement ratio and increasing the strength.

To prepare a grout test specimen, a firm, flat location should be selected where the mold can remain undisturbed for 48 hours. A 5/8-in.-thick pallet of wood 3-1/2 in. square is placed on the level surface with four masonry units typical of the project. Permeable paper, such as absorptive paper toweling, is taped to the face shell of each masonry unit and placed around the wood pallet to form the mold. The resulting mold is twice as high as it is wide (Fig. 2-13a). A layout giving a 3×3×6-in. mold is sometimes used. An alternate method of forming three prisms at one time is shown in Fig. 2-13b.

It should be noted that the paper towel lining

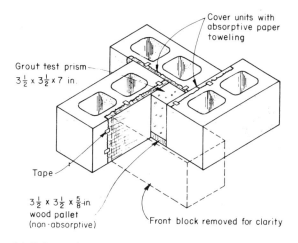

(a) Mold with four 8×8×16-in. block.

(b) Mold for forming three prisms at once.

Fig. 2-13. Alternate absorptive molds for grout test prisms.

prevents bond of grout to the masonry units while at the same time allowing absorption of the grout water by the units. Neatness in taping the paper to the units and care in consolidating the grout to avoid damage or wrinkling of the paper will prevent irregular-size test specimens with resultant varying test results.

The grout is poured into the mold in two layers. Each layer is rodded 25 times with a thin metal or wood puddling stick to eliminate entrapped air bubbles. The bottom layer is puddled throughout its depth, with the strokes distributed uniformly over the cross section of the mold. For the upper layer the stick is allowed to penetrate about 1/2 in. into the underlying layer. Grout may be vibrated if the corresponding section of the structure is also vibrated.

After the top of each prism is leveled with a trowel, the prisms are immediately covered with wet burlap or similar material to keep them damp. They are also protected against extreme changes in temperatures for 48 hours. Then the masonry units are removed and the prisms carefully packed in damp sawdust or sand for transport to the laboratory, where they receive additional moist-curing until tested.

The specimens are capped in accordance with the applicable provisions of Standard Method of Capping Cylindrical Concrete Specimens, ASTM C617. They should be tested in a damp condition in accordance with the applicable provisions of ASTM C39, Standard Method of Test for Compressive Strength of Cylindrical Concrete Specimens.

A wall does more than merely enclose a building in an attractive fashion. It must have strength to support floors and roofs and to resist the buffeting of nature. It must be a shield against noise, heat or cold, and fire damage. This chapter is intended to help architects, engineers, and builders design and construct concrete masonry walls that fill these needs.

Strength and Structural Stability

Modern concrete masonry wall construction is of two general types: plain and reinforced. These classifications are characterized by some differences in mortar type requirements, use of reinforcing steel, and erection techniques. Both types are usually subject to the provisions of applicable building codes.*

Plain concrete masonry is the ordinary, essentially unreinforced type that has been in use for many years. Any steel reinforcement used in this type of concrete masonry is generally of light gage and placed in relatively small quantities in the horizontal joints.

Reinforced concrete masonry is the type in which reinforcing steel is so placed and embedded that the masonry and steel act together in resisting forces. This structural behavior is obtained by placing deformed reinforcing steel bars in continuous vertical and horizontal cavities in the masonry and then filling these

*Partially reinforced masonry is recognized by Sec. 10.7 of the 1970 report of ACI Committee 531 (Ref. 52).

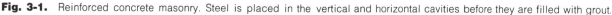

Fig. 3-1. Reinforced concrete masonry. Steel is placed in the vertical and horizontal cavities before they are filled with grout.

Fig. 3-2. High-rise load-bearing walls with round fluted concrete masonry units.

cavities with properly consolidated portland cement grout. Structural bond develops between the hardened grout, the bars, and the masonry units, permitting the theory of reinforced concrete design to be adapted to produce buildings of *engineered* concrete masonry.

Reinforced concrete masonry is used where the compressive, flexural, and shearing loads are higher than can be accommodated with plain concrete masonry. It is required by code in areas of recurring hurricane winds or earthquake activity where major damage to buildings is highly probable.

Structural Tests

Extensive testing of representative concrete masonry assemblages by recognized laboratories over a period of many years has established concrete masonry wall construction as a reliable and predictable structural system. Some of the test methods developed in structural research studies* of concrete masonry and other wall forms have been adopted as standards and are described in ASTM E72, Standard Methods of Conducting Strength Tests of Panels for Building Construction.

Analyses of accumulated test data have produced useful, reproducible relationships between the strengths of masonry components and the strengths of completed walls.** From these relationships building

---

*See Ref. 40.

**The strength of concrete masonry units also contributes structurally in floor and roof systems, as shown in Chapter 8.

code authorities have established allowable design stresses under various loading conditions or formulas for their calculation. Furthermore, structural research has enabled the development of quality control criteria to ensure fulfillment of design requirements in the completed structure.

Fig. 3-3. Concrete masonry wall being tested for flexural strength at the University of Illinois.

Fig. 3-4. A concrete masonry research project in the structural research laboratory of the Portland Cement Association. Specimens are being investigated for shrinkage and creep.

Design Methods

The model building codes in the United States and Canada and most local codes contain empirical values for allowable stresses and other factors relating to the design of concrete masonry walls. These conservative and somewhat arbitrary numbers have been carried over from the years when the structural behavior of masonry wall construction was not as well understood as it is today. However, some of these same building codes now permit the design of masonry walls by rational methods. A number of current codes and regulations for structural design are given in the reference list. A full explanation of the design methods is beyond the scope of this handbook.

It is necessary that the design of engineered concrete masonry be the responsibility of qualified engineers or architects. In such design, allowable compressive and flexural stresses are related to the ultimate net compressive strength of the masonry, f'_m, in pounds per square inch. The value used for this in design calculations may be determined by one of two methods:

1. A value of f'_m may be assumed from knowledge of what strength will result in an economical building design. Then the governing design code is consulted to determine the strength of block and the type of mortar required to achieve this value of f'_m. Fig. 3-5 shows typical design values.
2. The alternative method is to conduct compression tests in advance of design. Prisms are constructed as shown in Fig. 3-6, using the units and mortar contemplated for the job. If the structure is to have grouted cores, the cores of the prisms are filled with grout. No reinforcing bars are used in the prisms, but a two-wythe prism may contain metal ties. Prisms are cured in accordance with rules set forth in the code or specification. After the tests, the value of f'_m is calculated by dividing the ultimate test load by the solid cross-sectional area of the prism.

When prism strengths are specified, it must be remembered that the results will not be precisely uniform but will vary according to laws of probabilities. Some prisms will test higher than required, some lower. This phenomenon has not been codified for concrete masonry as it has been for concrete under ACI Standard 214, Recommended Practice for Evaluation of Compression Test Results of Field Concrete. Under that standard and ACI 318, Building Code Requirements for Reinforced Concrete, the test strength to be specified for construction must exceed the design strength by a certain amount, and an individual test value may fall below the design strength by a specified amount. To be realistic, some such procedure is necessary for concrete masonry. Values of f'_m determined from the first method are conserva-

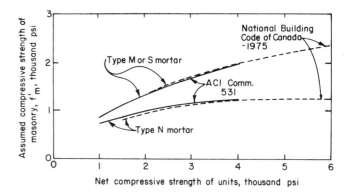

Fig. 3-5. Value of f'_m based on strength of individual concrete masonry units.

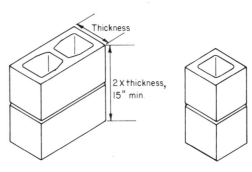

(a) Hollow concrete block prisms

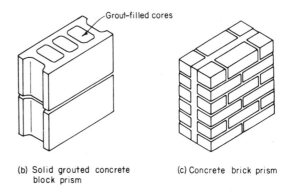

(b) Solid grouted concrete block prism

(c) Concrete brick prism

Fig. 3-6. Concrete masonry compression-test prisms.

tive, and values from the second method must be increased to allow for the scattering of test results.

Effect of Loads

Naturally, the various loads on a wall help determine how thick and how strong it must be. Vertical loads include the weights of floors, walls, and roof above, as well as the weight of the wall itself.* The example of allowable loading given in Table 3-1 is calculated for loading centered on a hollow concrete

Table 3-1. Allowable Axial Loads on Single-Wythe, Hollow Concrete Masonry Walls*

Height of wall (h), ft.	Allowable axial load, plf				
	t = 4	6	8	10	12
	t' = 1	1.19	1.25	1.50	1.50
6	5,593	7,309	7,833	9,454	9,448
7	5,112	7,142	7,740	9,384	9,382
8	4,477	6,923	7,632	9,301	9,314
9	3,664	6,648	7,497	9,206	9,238
10	2,651	6,309	7,335	9,090	9,154
11	1,420	5,900	7,146	8,959	9,057
12		5,416	6,927	8,807	8,952
13		4,844	6,663	8,636	8,829
14		4,181	6,369	8,445	8,703
15		3,426	6,030	8,227	8,555
16		2,561	5,646	7,978	8,396
17		1,584	5,220	7,703	8,219
18		493	4,740	7,400	8,028
19			4,200	7,067	7,812
20			3,615	6,699	7,584
21			2,964	6,291	7,336
22			2,253	5,847	7,062
23			1,476	5,367	6,770
24				4,848	6,452
25				4,285	6,110
26				3,679	5,746
27				3,023	5,350
28				2,324	4,932
29				1,570	4,488

*Adapted from Plate 4-16S of the Army-Air Force technical manual (Ref. 59), with

Allowable load $N = 0.20f'_m \left[1 - \left(\frac{12h}{40t} \right)^3 \right] A_n$

where t = nominal wall thickness, in.
 t' = width of face-shell mortar bed, in.
 f'_m = ultimate net prism strength, psi = 1,350
 A_n = net area of mortar bed, sq.in. per ft. = $24t'$

As required by Ref. 52, allowable heights in excess of h/t = 20 were determined by design calculations and are unshaded.

block wall braced at the top. If floor or roof joists do not rest on the center of the wall, the allowable loads are less.

It is assumed that, for all the allowable wall loads shown in the tables and graphs of this section, proper engineering or architectural supervision of construction (quality assurance) will take place. If this will not occur, the design codes specify that the loads must be drastically reduced. Inspection of engineered concrete masonry is necessary if the economic potential is to be realized.

Lateral loads induce significant stresses in a wall. Basement walls are subject to the lateral pressure of the earth. Also, if the earth is not drained by tile around the base of the foundation, hydrostatic pressure at the wall surface will cause greater flexural stresses in the wall. Table 3-2 shows required thicknesses of foundation walls subject to lateral earth pressure. Note that in two cases the wall does not have to be as thick with a masonry wall above as with a wood frame above; this is because the heavier superimposed load counteracts the tensile bending stresses to a greater extent. The use of reinforced masonry is an effective method of resisting large lateral pressures.

Wind and earthquakes also cause lateral pressure on exterior walls. Wind pressure is often given by the formula

$$p = 0.00256 \times FV^2$$

where p is the wind pressure in pounds per square foot; F, a correction factor for height; and V, the basic wind velocity in miles per hour. For ordinary rectangular

---

*Weights of concrete masonry walls with a single wythe (a continuous vertical section of masonry one unit in thickness) are given in Chapter 6, Table 6-1 (page 120). The weight of a reinforced, grouted wall can be determined by adding the average weight applicable from Table 6-1 to the weight of the grout, derived from Table 6-5 (page 122), and the weight of the reinforcing steel.

Table 3-2. Foundation Wall Design Criteria*

Foundation wall	Maximum height of unbalanced fill, ft.**	Minimum thickness, in.	
		Wood frame above	Masonry or masonry veneer above
Hollow concrete masonry	3	8	8
	5	8	8
	7	12	10
Solid concrete masonry	3	6	8
	5	8	8
	7	10	8

*Adapted from Sec. 601-16 of the HUD Manual of Acceptable Practices (Ref. 64).
**Height of finish grade above basement floor or inside grade.

buildings the wind pressure on the windward wall is 80% p and on the leeward wall the suction is about 50% p; gusts of wind add to these pressures.*

For wall resistance to earthquake forces, a typical building code provision in seismic regions calls for a lateral design load of 20% of the dead weight of the wall. Usually this will be less than the wind load on outside walls. For partitions and interior walls, codes often specify a 5-psf minimum load.

A concrete masonry wall will resist lateral loads in the horizontal and vertical spans, depending on its height and the length between cross walls. However, Fig. 3-7 shows that bending due to wind is principally in the vertical span. The diagram represents a typical one-story wall bounded by laterally supported, vertical control joints and having no windows or doors.

Although further refinements to the problem of wind distribution appear in other literature,** an example of allowable height of a one-story wall for wind, spanning only vertically, is given in Fig. 3-8. Fig. 3-9 gives allowable horizontal span of the same type of wall for wind, spanning only horizontally.

If a wall resists wind in its vertical span, a concentric compressive vertical load on the wall will increase its strength up to a point. For example, in Fig. 3-8, a 12-in. wall has an allowable height of 16 ft. against a wind pressure of 20 psf. However, Table 3-3 shows that, as the axial load increases on this same wall, its allowable height increases (up to a point). On the other

hand, when a wall is subject to a vertical load but resists wind by its horizontal span, the two effects are not considered to be additive.

As mentioned previously, reinforcing bars increase the strength of concrete masonry walls. This can be seen by comparing a reinforced wall in Table 3-4 with a similar but plain (nonreinforced) wall in Table 3-3.

Cavity walls have different strength characteristics than single-wythe walls. Wind pressure is resisted by non-load-bearing cavity walls, as shown in Figs. 3-10 and 3-11. These charts are for walls resisting wind in the vertical and horizontal spans, respectively.

Each wythe in a cavity wall helps resist wind by acting as a separate wall. If both wythes have the same thickness, they resist the wind equally but do not act

*See ANSI A58.1.
**See TEK 37, Ref. 9.

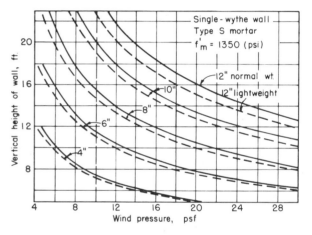

Fig. 3-8. Allowable height of non-load-bearing concrete masonry single-wythe walls for wind. (Adapted from Plate 4-11S, Ref. 59.)

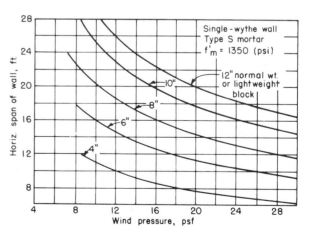

Fig. 3-9. Allowable horizontal span of non-load-bearing concrete masonry single-wythe walls for wind. (Adapted from Plate 4-12S, Ref. 59.)

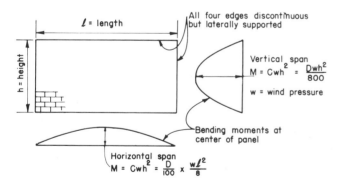

	Bending moment coefficient (C)		Virtual percent of wind creating bending (D)	
ℓ/h	Vertical span	Horizontal span	Vertical span	Horizontal span
1	0.044	0.044	35	35
1-1/2	0.078	0.043	62	15
2	0.100	0.037	80	7
2-1/2	0.112	0.032	90	4
3	0.118	0.029	95	3

Fig. 3-7. Distribution of wind forces on a wall panel.

Table 3-3. Allowable Heights of Single-Wythe, Hollow Concrete Masonry Walls for Concentric Loads and Wind*

| Axial load, plf | Allowable height of wall, ft. | | | | | | | | | | | | | | |
| | t = 4 | | | 6 | | | 8 | | | 10 | | | 12 | | |
	w = 10	20	30	10	20	30	10	20	30	10	20	30	10	20	30
0	7.54	5.22	4.22	11.61	7.91	6.35	15.53	10.40	8.29	20.33	13.41	10.77	24.96	16.09	12.61
500	8.65	6.85	5.45	13.10	9.80	7.80	17.35	12.25	10.10	21.85	16.00	12.50	27.50	18.50	14.70
1,000	9.12	7.87	6.38	13.70	11.25	9.08	18.05	14.42	11.57	22.65	17.67	14.10	28.25	20.87	16.52
1,500	9.35	8.50	7.20	14.05	12.40	10.15	18.55	16.00	12.90	23.15	19.40	15.40	28.62	22.80	18.10
2,000	9.48	8.85	7.85	14.21	12.90	11.15	18.80	16.75	14.00	23.45	20.65	16.70	28.80	24.50	19.60
2,500	9.48	8.95	8.15	14.25	13.15	11.90	18.95	17.15	15.10	23.65	21.30	18.00	28.90	25.30	21.25
3,000	9.35	8.95	8.30	14.20	13.30	12.35	18.97	17.32	15.65	23.70	21.65	19.15	28.85	25.70	22.40
3,500	9.10	8.85	8.30	14.02	13.32	12.55	18.85	17.40	15.95	23.65	21.85	19.90	28.75	25.95	23.05
4,000	9.65	8.55	8.15	13.73	13.22	12.60	18.65	17.35	16.05	23.52	21.95	20.30	28.45	26.05	23.40
4,500	8.00	8.00	7.85	13.29	13.00	12.50	18.20	17.20	16.05	23.25	21.90	20.47	28.10	26.05	23.65
5,000	7.25	7.25	7.25	12.65	12.55	12.25	17.60	16.90	15.92	22.90	21.85	20.55	27.50	25.95	23.72
5,500	6.30	6.30	6.30	11.80	11.75	11.70	16.50	16.35	15.62	22.35	21.60	20.45	26.65	25.65	23.65
6,000	—	—	4.80	10.75	10.75	10.75	15.20	15.20	15.10	21.55	21.15	20.25	25.50	25.15	23.45
6,500	—	—	—	9.50	9.50	9.50	13.50	13.50	13.50	20.55	20.45	19.85	24.05	24.05	23.05
7,000	—	—	—	7.85	7.85	7.70	11.60	11.60	11.70	19.30	19.30	19.20	22.45	22.45	22.20
7,500	—	—	—	—	—	—	9.50	9.50	9.50	17.65	17.65	17.65	20.45	20.45	20.45
8,000	—	—	—	—	—	—	3.10	3.10	3.10	15.90	15.90	15.90	18.10	18.10	18.10

*Adapted from Plate 4-25S of the Army-Air Force technical manual (Ref. 59), with *t* the nominal wall thickness, in., and *w* the wind load, psf. The mortar bed widths are the same as in Table 3-1. Mortar is Type S and f'_m = 1,350 psi. Design is based on Refs. 52 and 58. As required by Ref. 52, allowable heights in excess of $h/t = 20$ were determined by design calculations and are unshaded.

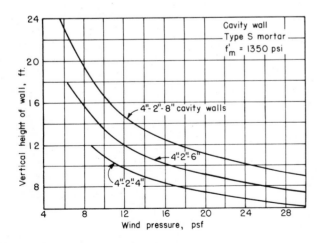

Fig. 3-10. Allowable height of non-load-bearing concrete masonry cavity walls for wind. (Adapted from Plate 6-4S, Ref. 59.)

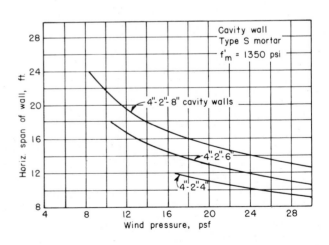

Fig. 3-11. Allowable horizontal span of non-load bearing concrete masonry cavity walls for wind. (Adapted from Plate 6-5S, Ref. 59.)

Table 3-4. Allowable Vertical Axial Loads of Single-Wythe, Hollow, Reinforced Concrete Masonry Walls for Wind*

Height, ft.	6-in. wall with No. 5 bars			8-in. wall with No. 6 bars			10-in. wall with No. 7 bars			12-in. wall with No. 8 bars		
	$w=10$	20	30	10	20	30	10	20	30	10	20	30
6	19.7	18.2	16.8	25.9	24.9	23.9				38.1	37.5	37.0
8	17.8	15.3	12.8	24.7	23.0	21.2	31.3	30.0	28.7			
10	15.4	11.7	8.0	23.1	20.4	17.8	30.1	28.1	26.1	36.6	35.0	33.4
12	12.3	7.6	2.8	21.0	17.4	13.7	28.7	25.8	23.0			
14	8.9	3.5		18.6	13.8	9.1	26.8	23.0	19.3	34.1	31.1	28.0
16				15.7	10.0	4.4	24.7	19.9	15.2			
18				12.5	6.2		22.2	16.5	10.8	30.5	25.7	21.0
20				9.2	2.7		19.4	12.9	6.3			
22							16.4	9.2	2.0	25.8	19.3	12.7
24							13.2	5.7				
26										20.2	12.2	4.3
30										14.0	5.5	

*Adapted from NCMA design tables (Ref. 20), with f'_m = 2,000 psi; w = wind load, psf. Vertical axial load and reinforcing bars are centered on the wall (the bars at 32 in. oc).

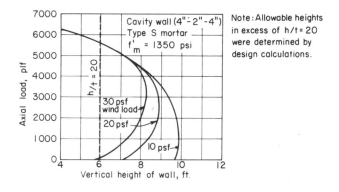

Fig. 3-12. Allowable load and vertical height of concrete masonry cavity walls for wind. Both wythes are 4 in. and vertical load is centered on the inside wythe. (Adapted from Plate 6-6S, Ref. 59.)

as if fully bonded together. For example, in Fig. 3-10 a 10-in. cavity wall (marked 4″-2″-4″) exposed to a 10-psf wind has an allowable height of 11 ft., but in Fig. 3-8 the 10-in. single-wythe wall of the same quality materials and exposed to a 10-psf wind has an allowable height of 20 ft. This illustrates that cavity walls have less strength. However, strength is often not a decid-

ing factor and cavity walls have distinct advantages: rain resistance and good insulation.

Vertical loads on cavity walls are often carried only by the inner wythe. Fig. 3-12 shows some allowable loads and heights for this condition where horizontal span is not involved.

Figs. 3-7 through 3-12 are based on the allowable stresses presented in Refs. 52 and 58. These charts are for hollow concrete masonry walls with the mortar bed widths given in Table 3-1; bending moments are computed for simple spans.

When vertical loads are not centered on the wall (any wall), high bending stresses are created. Their effects can be analyzed with structural engineering principles and then evaluated according to the allowable stresses given in the governing building code. Design tables* are available to facilitate this work. An example appears in Table 3-5. The design problem is greater if wind or earthquake stresses in the vertical span are also present, and this should be investigated by engineering analysis.

*See Refs. 12, 14, 20, and 59.

Table 3-5. Allowable Eccentric Loads on Hollow Concrete Masonry Walls*

e, in.	Allowable load, kips per ft., for 8-in. wall		e, in.	Allowable load, kips per ft., for 10-in. wall		e, in.	Allowable load, kips per ft., for 12-in. wall	
	Type M&S mortar	Type N mortar		Type M&S mortar	Type N mortar		Type M&S mortar	Type N mortar
2.00	2.74	1.90	2.25	3.97	2.76	3.00	4.60	3.20
2.20	2.11	1.47	2.50	2.95	2.05	3.25	3.59	2.50
2.40	1.72	1.19	2.75	2.35	1.63	3.50	2.99	2.08
2.60	1.45	1.01	3.00	1.95	1.36	3.75	2.56	1.78
2.80	1.25	0.87	3.25	1.67	1.16	4.00	2.23	1.55
3.00	1.10	0.77	3.50	1.46	1.01	4.25	1.98	1.38
3.20	0.98	0.68	3.75	1.29	0.90	4.50	1.78	1.24
3.40	0.89	0.62	4.00	1.16	0.81	4.75	1.61	1.12
3.60	0.81	0.56	4.25	1.06	0.73	5.00	1.47	1.03
3.80	0.74	0.52	4.50	0.97	0.67	5.25	1.36	0.95

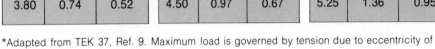

*Adapted from TEK 37, Ref. 9. Maximum load is governed by tension due to eccentricity of load, e, as shown above right.

Table 3-6. Required Reinforcement for Simply Supported Concrete Masonry Lintels*

Type of loading**	Nominal size, in., of lintel section	Reinforcing bars based on clear span							
		3 ft. 4 in.	4 ft. 0 in.	4 ft. 8 in.	5 ft. 4 in.	6 ft. 0 in.	6 ft. 8 in.	7 ft. 4 in.	8 ft. 0 in.
Wall load	6×8 6×16	1—No. 3 —	1—No. 4 —	1—No. 4 —	2—No. 4 —	2—No. 5 1—No. 4	— 1—No. 4	— 1—No. 4	— 1—No. 4
Floor and roof load	6×16	1—No. 4	1—No. 4	2—No. 3	1—No. 5	2—No. 4	2—No. 4	2—No. 5	2—No. 5
Wall load	8×8 8×16	1—No. 3 —	2—No. 3 —	2—No. 3 —	2—No. 4 —	2—No. 4 —	2—No. 5 —	2—No. 6 2—No. 5	— 2—No. 5
Floor and roof load	8×8 8×16	2—No. 4 2—No. 3	— 2—No. 3	— 2—No. 3	— 2—No. 4	— 2—No. 4	— 2—No. 4	— 2—No. 4	— 2—No. 5

*Adapted from TEK 25, Ref. 9. Design assumptions: bearing length each end = 8 in.; f'_m = 2,000 psi; wall load = 300 plf; floor and roof load (includes wall load) = 1,000 plf; weight of 8-in. lintel (included in wall load) = 50 plf; and weight of 16-in. lintel (included in wall load) = 100 plf. Lintel cross sections are shown below.
**Including weight of lintel.

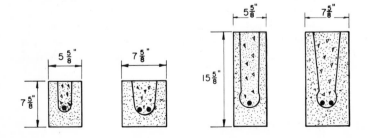

An easier problem for the designer is to see that, above each window or door opening in a wall, there is a structural beam or lintel to carry the wall loads, as shown in Fig. 3-13. Table 3-6 gives data for typical reinforced concrete masonry lintels. Lintel design is discussed further in the next chapter.

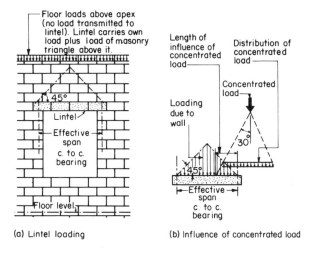

(a) Lintel loading   (b) Influence of concentrated load

Fig. 3-13. Loads supported by lintels.

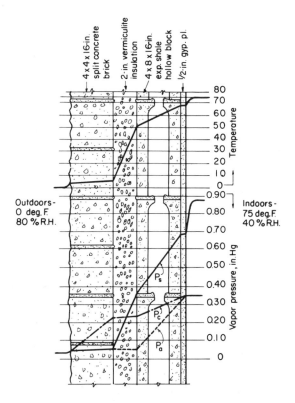

Fig. 3-14. Temperature and vapor pressure gradients for insulated concrete masonry cavity wall construction ($R = 7.26$, $U = 0.14$).

Thermal Insulation

In selecting and sizing the heating and cooling equipment for a building, the designer must first determine the heating and cooling loads involved. These loads are made up of various exchanges of heat, including transfer of heat through exterior walls.

When the outdoor temperature is below the established indoor temperature, heat is lost from the building and the heating system must replace it. Conversely, when the outdoor temperature is above the indoor temperature, heat is gained. In air-conditioned buildings this heat is absorbed by the cooling system. The upper part of Fig. 3-14 gives an example of heat flow through a concrete masonry wall. The temperature gradient varies from 75 deg. F. inside to 0 deg. F. outside.

U and R Values Defined

Essential to the designer's calculations regarding the flow of heat through walls as well as other building components are U values, the coefficients or indices of total heat flow rate. They express the total amount of heat in British thermal units (Btu) that 1 sq.ft. of wall (or ceiling or floor) will transmit per hour for each degree Fahrenheit of temperature difference between the air on the warm and cool sides.

Another index of heat transfer is the R value, which is a measure of the resistance that a building section, material, and air space or surface film offers to the flow of heat. The R value is the reciprocal of a heat transfer coefficient such as U or f. The R value is for a *stated* thickness of a building section, and the unit of value for R is $(°F \cdot h \cdot ft^2)/Btu$. R values are particularly useful for estimating the effect of components of a section on the total heat flow because they can be directly added.

The overall heat transmission coefficient of the wall, the U value, includes the effect of air film or surface conductance for the inside of the wall, f_i, as well as for its outside, f_o. A U value is calculated by taking the reciprocal of this total: $1/f_i + 1/f_o +$ the sum of the R values for each component of the wall.* U values cannot be added or subtracted with meaningful results.

The U value for a wall section can be determined by test (ASTM Method C236, Test for Thermal Conductance and Transmittance of Built-up Sections by Means of the Guarded Hot Box), and many wall tests have been conducted for this purpose over the years. Actually, the rational method for estimating the U value by calculation was devised because of the innumerable types and combinations of materials going into modern building walls and the continuing development of new

---

*See Refs. 28 and 29.

Table 3-7. Heat Resistance (R) Values of Single-Wythe Concrete Masonry Walls*

Nominal wall thickness, in.	Insulation in cells**	R value based on concrete unit weight				
		60 pcf	80 pcf	100 pcf	120 pcf	140 pcf
4	Filled	3.36	2.79	2.33	1.92	1.14
	Empty	2.07	1.68	1.40	1.17	0.77
6	Filled	5.59	4.59	3.72	2.95	1.59
	Empty	2.25	1.83	1.53	1.29	0.86
8	Filled	7.46	6.06	4.85	3.79	1.98
	Empty	2.30	2.12	1.75	1.46	0.98
10	Filled	9.35	7.45	5.92	4.59	2.35
	Empty	3.00	2.40	1.97	1.63	1.08
12	Filled	10.98	8.70	6.80	5.18	2.59
	Empty	3.29	2.62	2.14	1.81	1.16

*Adapted from TEK 38, Ref. 9. R values (defined in text) do *not* include the sums of the effect of air film or surface conductance on the inside of the walls ($1/f_i$ = 0.68) and on the outside ($1/f_o$ = 0.17).

**Loose-fill insulation such as perlite, vermiculite, or others of similar density.

types. The method is based on the heat-flow resistance concept and principles relating to the flow of electricity. Heat-flow resistance is analogous to electrical resistance. Thus, the flow of heat is directly proportional to the temperature difference and inversely proportional to the heat resistance.

Selecting the *U* Value

The hourly heat flow (Btu of heat loss or gain) through the exterior wall construction of a building can be determined by multiplying the *U* value of the section by the total net area in square feet (with areas of windows and doors deducted and dealt with separately) and then multiplying that answer by the design temperature difference between the inside and the outside air. Since the surface area and temperature difference are generally established or fixed quantities, the *U* value remains alone as subject to design adjustment.

In selecting the proper *U* value for a wall section, the designer will consider and weigh several factors. The relative importance of each will depend upon the type of building and its occupancy. For buildings intended for residential use—houses, apartments, dor-

Table 3-8. Heat Transmission Values of Some Typical Concrete Masonry Walls*

Components	Heat resistance values, R**							
	A	B	C	D1	D2	E	F1	F2
Surface film (outside)	0.17	0.17	0.17	0.17	0.17	0.17	0.17	0.17
8-in. hollow concrete masonry at 100-pcf density (cores open)	1.75				1.75	1.75		
Cores filled with bulk insulation		4.85					4.85	4.85
Concrete brick at 140-pcf density		0.44				0.44		
2-in. air cavity		0.97						
4-in. hollow concrete masonry at 100-pcf density		1.40				1.40		
2-in. bulk insulation in cavity						4.00		
Batt insulation between furring strips: 1 in.				3.70			3.70	
2 or 2-1/4 in.					7.00			7.00
1/2-in. gypsum board interior finish				0.45	0.45		0.45	0.45
5/8-in. plaster, lightweight aggregate		0.39				0.39		
Surface film (inside)	0.68	0.68	0.68	0.68	0.68	0.68	0.68	0.68
TOTAL resistance, R	2.60	4.05	5.70	6.75	10.05	7.08	9.85	13.15
U value, 1/R	0.384	0.247	0.175	0.148	0.100	0.141	0.102	0.076

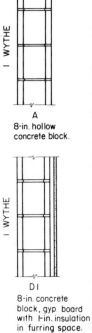

A
8-in. hollow concrete block.

D1
8-in. concrete block, gyp. board with 1-in. insulation in furring space.
D2
2-in. insulation used.

*Adapted from TEK 38, Ref. 9. The overall heat transmission coefficients, called *U* values, are determined by the heat resistance values, R, as explained in the text.
**Key for wall design shown at right.

mitories, hotels, etc.—human comfort needs probably will govern the selection. For commercial buildings devoted to manufacturing operations, process control requirements may be the primary factor in deciding the *U* value. In all cases, operational costs, i.e. the cost of fuel or electric power, will require careful consideration.

The requirements for building insulation have been undergoing a thorough change in recent times due to the emphasis on conservation of energy and fuels for winter heating and summer cooling. Since national standards will probably continue to evolve for several more years, no attempt will be made here to quote insulation requirements of any regulatory agencies. The reader is referred to releases from such organizations as the local building department, U.S. Depart-

ment of Housing and Urban Development, National Research Council of Canada (especially its Code for Residential Construction), and American Society of Heating, Refrigerating and Air-Conditioning Engineers, Inc.

Typical Values for Concrete Masonry Walls

Concrete masonry walls offer insulation qualities combined with architectural appeal. Table 3-7 lists *R* values for single-wythe walls and Table 3-8 gives *R* values for other common types of wall components. In the latter table the *U* value for each complete assembly (bottom line) was computed after the *R* value of

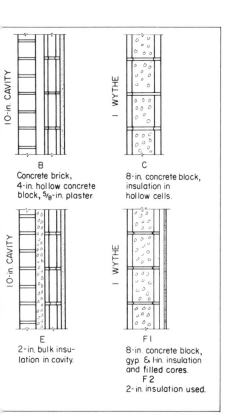

B
Concrete brick, 4-in. hollow concrete block, 5/8-in. plaster.

C
8-in. concrete block, insulation in hollow cells.

E
2-in. bulk insulation in cavity.

F1
8-in. concrete block, gyp. & 1-in. insulation and filled cores.
F2
2-in. insulation used.

Table 3-9. Estimated *U* Values for 8-In. Hollow Concrete Masonry Walls*

Wall details	60 pcf	80 pcf	100 pcf	120 pcf	140 pcf
No insulation	0.32	0.34	0.38	0.43	0.55
No insulation, 1/2-in. gypsum board on furring strips	0.21	0.23	0.25	0.27	0.31
No insulation, 1/2-in. foil-backed gypsum board on furring strips	0.15	0.15	0.16	0.17	0.19
Loose-fill insulation in cores	0.12	0.14	0.18	0.21	0.35
Loose fill in cores, 1/2-in. gypsum board on furring	0.10	0.12	0.14	0.17	0.24
Loose fill in cores, 1/2-in. foil-backed gypsum board, furring	0.08	0.10	0.11	0.12	0.16
1-in. rigid glass fiber, 1/2-in. gypsum board	0.14	0.14	0.15	0.15	0.17
1-in. polystyrene, 1/2-in. gypsum board	0.12	0.12	0.12	0.13	0.14
1-in. polyurethane, 1/2-in. gypsum board	0.10	0.10	0.11	0.11	0.12
Loose fill in cores plus 1-in. rigid glass, 1/2-in. gypsum board	0.08	0.09	0.10	0.11	0.14
Loose fill in cores plus 1-in. polystyrene, 1/2-in. gypsum board	0.07	0.08	0.09	0.10	0.12
Loose fill in cores plus 1-in. polyurethane, 1/2-in. gypsum board	0.07	0.07	0.08	0.09	0.11
R-7 blanket insulation, 1/2-in. gypsum board, furring	0.09	0.10	0.10	0.10	0.11

Header: *U* value based on density of concrete used in block

*Adapted from TEK 67, Ref. 9 *U* values of other wall types and sizes are also given in TEK 67.

each component was tabulated and then added to give the total *R* value of each wall. In Table 3-9 the *U* values of 8-in. walls with various combinations of finish and insulation are given directly.*

It is important to select the correct wall as insulation against the weather, but this should not obscure some larger considerations affecting the total cost of heating and cooling a building. Heat flow through a single pane of glass in winter is 1.13 Btu per square foot per degree Fahrenheit per hour. This is six or seven times the heat flow through a lightweight concrete block wall having filled cores.

In general, the heat flow through the walls of a building is only a small part of the total heat flow. For instance, it has been found** that:

	Total heat loss, percent
Low-rise apartment walls	4 to 16
High-rise apartment walls	5 to 32
Single-family residence walls	10 to 26
One-story industrial walls	13 to 54

Heat losses through roofs, floors, and fenestration are other important considerations, as are losses due to infiltration, exhausts, fresh air intakes, and opening of doors.

Effect of Moisture Content

Heat transfer (conductivity) values increase in walls as the moisture content increases. The relationships found† between the conductivity and the density and moisture content of concrete, mortar, or grout are shown in Fig. 3-15. The heat transfer values for oven-dry weights of concrete are frequently used. However, in an occupied building, the amount of moisture in the concrete or block may be termed "normally dry." This condition describes the equilibri-

um moisture content of concrete after extended exposure to 78 deg. F. air at 35 to 50% relative humidity. Note that conductivity in the normally dry condition is only slightly greater than in the oven-dry condition.

In the event that concrete masonry should become saturated with water, heat transfer will increase markedly. This condition could occur in a basement wall that is not dampproofed. In walls above grade the condition might prevail for a short time after a heavy driving rain.

A situation to avoid is having a sealer applied to the cold side of a wall. Moisture travels towards that side and, if its surface is sealed, the wall can approach saturation there, thus losing much of its insulating value. This illustrates the importance of using exterior paints that can "breathe," as discussed in Chapter 7.

Heat Gain

The same heat transfer coefficients are used in both heating and cooling design considerations, except that slight differences are made in the heat losses of the wall surfaces themselves. In addition, the calculation of cooling load may make use of temperature differentials that account for the effects of solar radiation and exterior surface reflectances as well as the mass or weight of the building parts.

Heat gain from solar radiation is affected by the orientation of the building and the reflectivity of the exterior surfaces. Light-colored surfaces will reflect the sun's rays much more effectively than dark-colored surfaces. This explains the wider use of white or light-shaded paints and finishes for wall surfaces in the southern regions of the United States.

Steady State Versus Dynamic Thermal Response

A distinction must be made between masonry and nonmasonry construction because of the difference in heat storage capacity. Heavy construction such as concrete masonry does not respond to temperature fluctuations as rapidly as does lightweight construction, even though the *U* values may be identical. Due to a "fly-wheel effect" the net transfer of heat through a concrete masonry wall section for a certain time period might actually be less. This cyclic phenomenon is termed "dynamic thermal response."

The practice in calculating heat transfer as outlined previously is based on a "steady-state" temperature differential or a constant difference between extreme

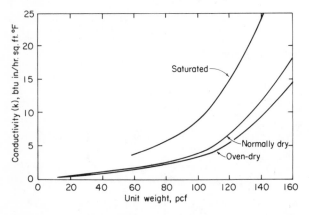

Fig. 3-15. Conductivity of concrete, mortar, or grout as affected by density and moisture. Aggregate type influences conductivity only as it affects the resulting unit weight and moisture content.

---

*See Ref. 49 for insulating values of other lightweight-aggregate concrete masonry walls.
**See Ref. 32.
†See Ref. 30.

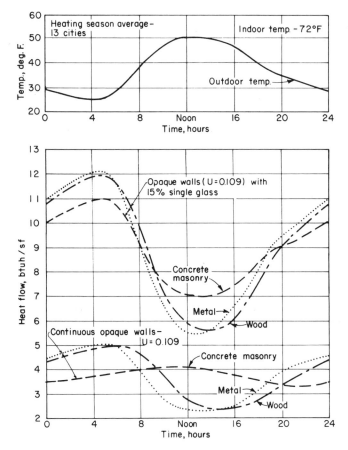

Fig. 3-16. Heat loss through masonry and nonmasonry walls, *U* values being equal.

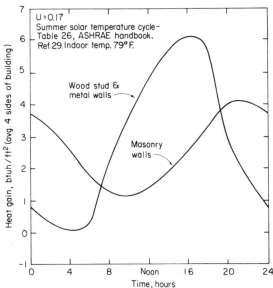

Fig. 3-17. Heat gain through masonry and nonmasonry walls, *U* values being equal.

outdoor and indoor air temperatures. The size of heating or cooling equipment is calculated on this basis, but higher *U* values are permitted in buildings with heavy walls, floors, and roofs because of the long-known fact that they act as heat reservoirs. This is of importance since the actual temperature differential between indoors and outdoors is not constant but fluctuates with the time of day.

The steady-state technique has been the accepted method of design for many years, although the theory and basic mathematics of cyclic temperature heat losses have been known for more than a century and a half. The complexity and expense of the calculations prevented engineers from using the sophisticated approach until the era of the electronic computer arrived.

To analyze the effect of daily temperature changes, a computer program for dynamic thermal response was developed by the National Bureau of Standards.* The computer program was verified with a structure built of 8-in. solid lightweight concrete masonry walls and a 4-in.-thick concrete roof. After the experimental structure was subjected to an outside cyclic temperature variation of 60 deg. F., it was confirmed that the heat flow rates calculated by the steady-state method

ranged from 32 to 69% higher than the rates of the dynamic thermal response program.

The NBS computer program was used in a PCA study** to compare heat losses of walls having equal *U* values. Fig. 3-16 depicts the winter heat losses of three types of walls subject to typical outdoor temperature changes. The times at which peak loads occur are different, and there is a significant difference in the peak rate of heat loss, the value traditionally used to size heating equipment.

Fig. 3-17 shows summer heat gain curves for two types of walls with equal *U* value. It illustrates the well-known fact that during a hot day a masonry building is cooler than a lighter building. The relative air-conditioning loads can be visualized for the two types of walls.

Control of Water Vapor Condensation

When warm, humid air is chilled to a certain point, as by a cold surface, condensation takes place (Fig. 3-18). To prevent condensation or sweating on the interior or room-side surface of a wall, the overall resistance to heat transmission of the wall must be such that the surface temperature will not fall below the dew (condensation) point of the room air. Dew points for various room temperatures and relative humidities are listed in Table 3-10. It shows, for example, that with a room temperature of 70 deg. F.

*See Ref. 31.
**See Ref. 37.

Fig. 3-18. Beads of water on the outside of this tumbler of ice water are caused by condensation of water vapor in the air as it comes in contact with the cold surface.

whether the wall has an overall high or low resistance to heat flow. Water vapor is a gas in the air and will diffuse through materials of building construction at rates that depend upon the water vapor permeability of the materials and the existing vapor pressure differential. The passage of water vapor through a material is not harmful in itself. It becomes of consequence only when, at some point along the water vapor flow path, a temperature level drops below the dew point and condensation occurs.

Excessive accumulation of condensed water vapor in concrete masonry may lead to efflorescence or temporary formation of frost within the wall. In most concrete masonry buildings freezing does not constitute a problem due to the daily fluctuations in temperatures. Frost thaws and is released through the outer wythe as vapor or through weepholes as condensate.

The risk of damage from frost buildup in walls is greatest during extended periods of very low temperatures outdoors and relative humidities exceeding about 50% indoors. However, when outdoor temperatures are very low, high relative humidities indoors are rare except in laundries, bathing areas, etc. The maximum humidity that can be continuously maintained in a room is limited by the dew point of the wall or window or any sensitive zones on them.

To minimize condensation and yet provide for human health and comfort, some agencies* recommend that the maximum relative humidity maintained indoors should be approximately as follows:

Outdoor temperature, deg. F.	Indoor relative humidity, percent
0	25
10	30
20	35
30	40

*See Ref. 41.

and relative humidity of 40%, the wall surface temperature should not fall below 45 deg. F. if sweating is to be avoided.

Normally, sweating of the interior surfaces of building walls is not a problem when proper attention is given to the insulating quality of the walls and the relative humidity is controlled within reasonable limits. Water vapor condensation that occurs on uninsulated basement walls below grade during humid periods of the summer or during basement laundering activity may require mechanical dehumidification or ventilation for its control.

The control of condensation within wall spaces and wall materials is more complex because it can occur

Table 3-10. Dew Point Temperatures

Dry bulb or room temperature, deg.F.	Dew point, deg.F., based on relative humidity									
	10%	20%	30%	40%	50%	60%	70%	80%	90%	100%
40	−9	5	13	19	24	28	31	34	37	40
45	−5	9	17	23	28	32	36	39	42	45
50	−1	13	21	27	32	37	41	44	47	50
55	3	17	25	31	37	41	45	49	52	55
60	6	20	29	36	41	46	50	54	57	60
65	10	24	33	40	46	51	55	58	62	65
70	13	28	37	45	51	56	60	63	67	70
75	17	31	42	49	55	60	65	68	72	75
80	20	36	46	54	60	65	69	73	77	80
85	23	40	50	58	65	70	74	78	82	85
90	27	44	55	62	69	74	79	82	86	90

An example of vapor pressure gradients appears in the lower part of Fig. 3-14 (page 51) for a steady-state condition in which moisture and heat are migrating to the outdoors. The vapor pressure is expressed in inches of mercury (Hg). Gradient P_s assumes saturated air (100% relative humidity) indoors, and gradient P_c assumes continuous vapor flow for the actual humidities but neglects the possibility of condensation. Condensation would occur at the point where gradients P_s and P_c intersect. Since that point is in the granular vermiculite fill where resistance to vapor flow is very low, the condensation would probably take place on the inside of the outer wythe. Then, since the temperature there is about 5 deg. F., frost would develop. Gradient P_a shows actual vapor pressure gradient.*

Generally when the designer is confronted with building service conditions conducive to water vapor condensation, he provides a vapor barrier on or as close as possible to the warm surface of the wall. The vapor barrier is a material—such as plastic film, asphalt-treated paper, and alumimum or copper foil—that will transmit not more than one grain of water vapor per square foot of surface (normal to the vapor flow path) per hour under a vapor pressure differential of 1 in. of mercury. Thus, the vapor barrier will reduce to a minimum the entrance of water vapor into the wall. Then the reduced amounts of water vapor penetrating the vapor barrier will pass through the outer layers of the wall by diffusion. Of course, any outer surface treatment such as paint must be of a type that will "breathe."

For a vapor barrier to be fully effective, it must be applied as a leakproof, continuous layer. Openings such as those provided by electrical outlet boxes must be given close attention, and avenues for vapor leakage through or around window and door frames must be stopped.

Acoustics

Noise is sound that is not wanted, and what is considered noise depends upon the individual and his (her) level of tolerance. People value their privacy and do not care to hear the movements of their neighbors. Homo sapiens is jealous of his (her) territorial rights, including invasion by sound (noise).

Acoustics, as a science and technology, is well advanced. The subject is complex, but if a few simple concepts are learned, the individual can do a great deal to assess acoustical problems. Good acoustical design in residences and offices can be achieved with absorptive surfaces and relatively heavy wall and floor construction. Concrete masonry is an excellent sound barrier because of its density.

*See Refs. 28 and 29.

Elements of Sound

The two important parameters used for the study of acoustics are frequency and decibels—or, loosely speaking, tone and loudness. The tone of a sound depends on the number of vibrations per second, the frequency. One vibration per second or one cycle per second (1 cps) is called a hertz, abbreviated Hz. On the piano, middle C has a frequency of 262 Hz. The lowest of the 88 keys on a piano has a frequency of 27 Hz and the highest a frequency of 4186 Hz. This is essentially the range of tones used in the study of building acoustics.

Pressure of sound is measured by the decibel. Each increase of 20 db (decibels) indicates a tenfold increase of pressure. However, the ear mechanism automatically reduces its sensitivity as the pressure increases. An increase of 10 db is a threefold increase in pressure but the loudness sensation to the ear is only doubled. Since the human ear has a greater range of sensitivity to intensity in the middle range of frequencies, loudness is determined by the pressure level, decibels, at a frequency of 1,000 cps (1000 Hz).

An idea of the decibel as a unit of measure of sound intensity is obtained from the following list:

Decibels	Sound
140	F84 jet at takeoff 83 ft. from tail
130	Threshold of discomfort
120	Thunder
110	Car horn at 3 ft.
100	Wood saw at 3 ft.
90	Noise inside a city bus
80	Noisy office or vacuum cleaner
70	Boeing 707 jet landing (at 3,300 ft.)
60	Average office
50	Average conversation
40	Quiet radio
30	Quiet conversation
20	Whisper at 4 ft.
10	Normal breathing
0	Threshold of audibility

Criteria for Acoustic Ratings

Building codes regulate the amount of noise that must be stopped by walls, floors, and ceilings. Typical values are in the range of 40 to 55 db of sound loss for airborne as well as impact sounds.

There are three principal types of ratings, as shown in Table 3-11. Each type may be identified at any individual frequency or by class. SAC and NRC values are in sabins, which are units measuring energy absorption and not sound or loudness. All other ratings are in decibels. The ratings have one thing in common: *the larger the number, the better the sound insulating quality of the wall or floor.*

Table 3-11. Acoustic Ratings

Type of rating	Rating by individual frequencies	Rating by class
Sound absorption (airborne)	SAC	NRC
Sound transmission loss (airborne)	STL	STC
Impact noise isolation or Impact sound isolation	ISPL	INR IIC

ASTM C423 Standard Test Method for Sound Absorption and Sound Absorption Coefficients by the Reverberation Room Method

ASTM C634 Standard Definitions of Terms Relating to Environmental Acoustics

ASTM E90 Standard Method for Laboratory Measurement of Airborne Sound Transmission Loss of Building Partitions

ASTM E336 Standard Test Method for Measurement of Airborne Sound Insulation in Buildings

ASTM E413 Standard Classification for Determination of Sound Transmission Class

ASTM E492 Standard Method of Laboratory Measurement of Impact Sound Transmission Through Floor-Ceiling Assemblies Using the Tapping Machine

ASTM E597 Tentative Recommended Practice for Determining a Single-Number Rating of Airborne Sound Isolation in Multiunit Building Specifications

Fig. 3-19. American Society for Testing and Materials (ASTM) standards for acoustics.

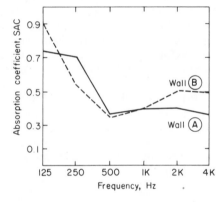

Fig. 3-20. Sound absorption test data for two 8-in. concrete masonry walls made with shale aggregate.

"Sound absorption" refers to the amount of airborne sound energy (sabins) absorbed on the wall surface adjacent to the sound. "Sound transmission loss" is the total amount of airborne sound (decibels) lost as it travels from one side of the wall or floor to the other. "Impact noise isolation" and "impact sound isolation" refer to the number of decibels lost through a floor from standardized impacts on top of the floor (or floor covering). ISPL, INR, and IIC ratings are not used for walls.

The ratings are governed by the ASTM standards given in Fig. 3-19. These standards are in general use in the United States and Canada.

Sound Absorption

Fig. 3-20 shows data for 8-in. lightweight concrete masonry walls tested in two different laboratories.

Note how the absorption varies with the frequency. The sound absorption coefficient (SAC) is the amount of sound energy absorbed compared to a perfectly absorptive surface such as an open window. Translated into decibels, SAC at 0.9 equals only 10 db; at 0.7, 5 db; at 0.5, 3 db; and at 0.3, 1.5 db. This is true regardless of the sound level. As a practical matter, it is difficult to lose much more than about 5 db of sound by absorption (reduce the loudness by 33%).

The noise reduction coefficient (NRC) is found by averaging the SAC values at frequencies of 250, 500, 1000, and 2000 Hz. In Fig. 3-20, Wall A has an NRC value of 0.46 (about 3 db).

Typical noise reduction coefficients are given in Table 3-12. However, a concrete block with uniformly fine texture, as shown at the top of Fig. 1-25 (page 23), may have an NRC value as high as 0.68 (5 db). In contrast, glass, plaster, and other smooth surfaces have NRC values of less than 0.05 (less than one-sixth of a decibel). Also note in Table 3-12 that *painted* concrete masonry has a reduced NRC value.

Sound Transmission Loss

If walls have sealed surface pores on at least one side, the sound transmission loss (STL) is related to the weight of the wall. When the pores are sealed, the wall tends to transmit sound by acting as a diaphragm, literally stopping sound from passing through it. Because of this, a heavier wall tends to reduce the transmission of sound more than a light wall, following what is known as "mass law." It holds true for homogeneous partitions that are nonporous and have uniform physical properties throughout the entire wall panel.

Table 3-13 shows the STC values for various types of concrete masonry walls. An example of STL test data for several concrete masonry walls is given in Fig. 3-21.

All of the stated values for sound transmission loss through a wall are meaningless if the wall panel has an opening. Large (as a window opening) or small (as an opening for a water pipe or electrical conduit), an opening can seriously alter the sound reduction capability of a wall. If maximum sound reduction is desired, any opening should be carefully avoided. For example, placing electrical outlet boxes back to back through a concrete masonry wall will provide an opening or flanking path through which sound can travel. Just as undesirable are any openings around doors, pipes, and air ducts. Another flanking path is provided by a partition that extends only to a suspended ceiling rather than to the floor or roof above. Calking is necessary where walls join and between walls and ceilings.

The sound level in a room is known as the masking sound or background noise; it fairly well swallows up lesser sounds transmitted through walls and floors.

Table 3-12. Approximate Noise Reduction Coefficients*

Material	Surface texture	Approximate NRC
Lightweight aggregate block, unpainted	Coarse	0.40
	Medium	0.45
	Fine	0.50
Heavy aggregate block, unpainted	Coarse	0.26
	Medium	0.27
	Fine	0.28
Deduct from above for painting		

Paint	Application	For 1 coat	For 2 coats
All	Spray	10%	20%
Oil	Brush	20%	55%
Latex	Brush	30%	55%
Cement	Brush	60%	90%

*Adapted from TEK 18, Ref. 9.

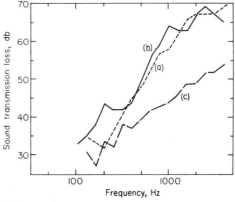

Notes:
Wall (a) is 8-in. hollow heavyweight concrete masonry, one side painted and the other side furred, plus 1/2-in. gypsum wallboard. Weight is 45.6 psf and STC is 49.
Wall (b) is 6-in. hollow lightweight concrete masonry, one side painted and the other side 1/2-in. gypsum board on resilient channel furring. Weight is approximately 26 psf and STC is 53.
Wall (c) is same as (b) but without finish. Weight is about 23 psf and STC is 44.

Fig. 3-21. Sound transmission loss test data for several concrete masonry walls.

This is because the decibel level of separate sounds are not directly additive. Table 3-14 shows how sound transmission loss through a wall panel affects noise conditions. The values are based on an assumed background noise level that corresponds to average conditions in most residences. Background noise levels in almost all offices and most institutions are higher.

Table 3-13. Data from Sound Transmission Loss Tests (ASTM E90-70) of Concrete Masonry Walls*

Wall description	Test No.**	Wall weight, psf	STC
Unpainted walls:			
8-in. hollow lightweight-aggregate units, fully grouted, No. 5 vertical bars at approx. 40 in. oc	1023-1-71	73	48
8-in. hollow lightweight-aggregate units	1144-2-71	43	49
8-in. composite wall—4-in. brick, 4-in. lightweight hollow units	1023-4-71	58	51
8-in. dense-aggregate hollow units	1144-3-71	53	52
10-in. cavity wall—4-in. brick, 4-in. lightweight hollow units	1023-6-71	56	54
Walls painted on both sides with 2 coats of latex paint:			
4-in. hollow lightweight-aggregate units	1379-5-72	22	43
4-in. hollow dense-aggregate units	1379-3-72	29	44
6-in. hollow lightweight-aggregate units	933-2-70	28	46
6-in. hollow dense-aggregate units	1379-1-72	39	48
8-in. hollow lightweight-aggregate units, fully grouted, No. 5 vertical bars at approx. 40 in. oc	1023-2-71	73	55
Walls plastered with 1/2-in. gypsum plaster on both sides:†			
8-in. composite wall—4-in. brick, 4-in. lightweight hollow units	1023-10-71	61	53
8-in. hollow lightweight-aggregate units, fully grouted, No. 5 vertical bars at approx. 40 in. oc	1023-9-71	79	56
10-in. cavity wall—4-in. brick, 4-in. lightweight hollow units	1023-8-71	59	57
Walls covered with 1/2-in. gypsum board on resilient channels:†			
4-in. hollow lightweight-aggregate units	1379-4-72	26	47
4-in. hollow dense-aggregate units	1379-2-72	32	48
8-in. composite wall—4-in. brick, 4-in. lightweight hollow units	1023-5-71	60	56
8-in. hollow lightweight-aggregate units	933-1-70	40	56
10-in. cavity wall—4-in. brick, 4-in. lightweight hollow units	1023-7-71	58	59
8-in. hollow lightweight-aggregate units, fully grouted, No. 5 vertical bars at approx. 40 in. oc	1023-3-71	77	60

*Adapted from Table 2, TEK 51, Ref. 9.
**Kodaras Acoustical Laboratories, Elmhurst, N.Y.
†Surface treatment on block side only of composite and cavity walls.

Impact Sound Isolation

Impact sound ratings were adopted to combat annoying sounds of footsteps and objects dropping above. For an effective sound barrier, a floor should have an impact sound isolation (IIC) rating of at least 40 db.

Although test data for floors made with concrete masonry are meager, other types of concrete floors have been tested many times. For example, here are the IIC ratings of several bare concrete floors:

	Decibels
4-in.-thick solid slab, 53 psf.	25
6-in.-thick hollow prestressed slab, 43 psf	23
8-in.-thick hollow slab, 45-60 psf	26
10-in.-thick solid slab, 121 psf	29

Table 3-14. Relationship Between Sound Transmission Loss Through a Wall and Hearing Conditions on Quiet Side*

Transmission loss, db	Hearing condition	Rating
30 or less	Normal speech can be understood quite easily and distinctly through the wall.	Poor
30 to 35	Loud speech can be understood fairly well. Normal speech can be heard but not easily understood.	Fair
35 to 40	Loud speech can be heard but is not easily intelligible. Normal speech can be heard only faintly, if at all.	Good
40 to 45	Loud speech can be faintly heard but not understood. Normal speech is inaudible.	Very good—recommended for dividing walls between apartments.
45 or more	Very loud sounds such as loud singing, brass musical instruments, or a radio at full volume can be heard only faintly or not at all.	Excellent—recommended for band rooms, music practice rooms, radio and sound studios.

*This table is based on the assumption that a noise corresponding to 30 db is continuously present on the listening side.

Table 3-15. Acoustic Rating Improvements for Concrete Floors with Various Treatments

Treatment	INR or IIC (impact) improvement, db
Carpet, pad, and acoustic ceiling	58
Carpet and pad	48
Acoustic ceiling	27
1/2-in. T&G wood parquet	25
Cork tile and furred ceiling	21
Plaster or gypsum board ceiling	8
Vinyl tile	4
2-in. concrete topping	0

It can be expected that concrete masonry will perform slightly better than concrete. In either case the sound isolation of the bare floor is marginal, but floor coverings and ceiling treatments are immensely helpful, as shown in Table 3-15. Floor or ceiling treatments that add at least 20 db to the IIC rating will solve the acoustical problem of impact noises on concrete floors.

Other Acoustical Considerations

Noise-producing equipment should be kept as far as possible from occupied areas, especially bedrooms. Flexible connectors should be used to couple mechanical equipment to pipes, ducts, and electric power. Also, pipes and ducts should not be firmly connected to parts of the building that can serve as sounding boards. Instead they should be supported by resilient connections to solid supports. Where they pass through walls and floors, they should be isolated from the construction by gaskets.

If a sound source must be close to a work area, sound barriers and sound absorbers around the source should be considered. In many instances quieter appliances and other equipment are the best solution. Manufacturers have become aware of the need and many "sound-rated" devices are now available, frequently at little additional cost.

When a single wall is used as a sound barrier, it is sometimes desirable to provide a resilient connection between the wall and the building frame. A 5- to 7-db improvement can result. Double walls perform better than single walls when of equal weight. Increased separation and sound-absorbing material in the cavity add to the performance of cavity walls. The sound transmission loss of a cavity wall is frequently about 8

Fig. 3-22. Slotted concrete block improve the acoustics.

Fig. 3-23. Traffic noise from expressways can be effectively shielded by a concrete masonry wall.

Table 3-16. Typical Noise Levels of Transportation Vehicles

Type of traffic	Noise level, db, based on distance from vehicle		
	50 ft.	200 ft.	800 ft.
Passenger car, 50-60 mph	70	58	46
Truck, max. highway speed	88	76	64
Diesel train, 30-50 mph	97	85	73

db better than a solid wall of equal weight and, if the two wythes are of unequal weight, as much as 4 db more can be added to the STC rating.

Selective absorption that matches the frequency of an unwanted sound is obtained by slotting or drilling the face shells of concrete masonry units, as shown in Fig. 1-22a (page 21). Fig. 3-22 here shows the result: an attractive and sound-absorbing wall.

Highway and Railway Sound Barriers

Those who live or work along heavy traffic routes are well aware of the din of traffic, typified in Table 3-16. Concrete masonry walls are helpful in reducing this noise level and at the same time provide a measure of privacy.

Studies in California* have shown that a 15-db sound reduction can readily be achieved by a concrete masonry wall. The level of sound reduction depends on the relative distances of the wall to the highway (or rail line) and to the listener as well as the height of the wall above the line of sight to the vehicle.

Fire Resistance

Firesafety is a major consideration in most building codes. The governing philosophy behind various code requirements with respect to fire is structural safety during the fire and containment of the fire. If a fire occurs, the construction should function to prevent its further spread. Thus, the components of buildings must have certain degrees of resistance as barriers to

*See Ref. 35.

the spread of fire, depending upon the nature of the occupancy, the fire zone classification, the character and concentration of combustibles, and other considerations.

The fire containment philosophy is perhaps best exemplified by the provisions for firewalls. Code requirements for firewalls are considerably more rigorous than for other walls and partitions. Firewalls must be capable not only of withstanding the heat effects of the most severe fire possible for the particular occupancy, but also of resisting the overturning and impact forces that may develop on either side from collapsing floors or other falling members. To ensure their lateral stability, firewalls are of noncombustible construction and may be specified to have a 4-hour resistance to heat transmission rather than a bare minimum of perhaps two hours. Concrete masonry is particularly well suited to firewall construction and a fire resistance of four or more hours can be readily achieved.

Evaluating Fire Resistance of Walls

Fire resistance is generally expressed in terms of the time that a sample of construction will withstand the standard fire test while still performing its purpose. The standard fire test is described in ASTM E119, Methods of Fire Tests of Building Construction and Materials, or CSA B54.3, Methods of Fire Tests of Walls, Partitions, Floors, Roofs, Ceilings, Columns, Beams and Girders. The extent and severity of the fire in the test furnace must conform to the standard time-temperature relationship shown in Fig. 3-25. This curve simulates the development of a severe fire.

Three end-point criteria are applied during the course of a standard fire test of a wall or partition:

1. Structural failure due to the effects of fire while carrying the safe design load.
2. Transmission of heat sufficient to cause an average rise in temperature of the unexposed surface (average for nine locations) of 250 deg. F. or a rise of 325 deg. F. at any one location. These temperature increases consider the possible spread of fire by ignition of combustibles placed near or against the wall surface away from the fire.
3. Passage of flame or heated gases hot enough to ignite cotton waste, or passage of water from the hose stream. After cooling, but within 72 hours, the wall must sustain twice the safe superimposed design load.

Any one of these criteria can be decisive should it occur first.

As part of the fire test, a hose stream test is required for any wall designed to have an endurance of one hour or more. The intent of this test is to determine the resistance of the construction to severe thermal shock,

Fig. 3-24. Fire resistance is but one of the many positive qualities of concrete masonry, giving a feeling of reassurance.

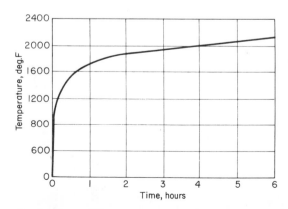

Fig. 3-25. Standard time-temperature curve for fire tests.

impact, and erosion caused by the water stream from a full-sized fire department hose. This test is specified to be made on a duplicate specimen exposed to the fire for at least half the time of the first specimen but not necessarily more than an hour. However, because of the stability of concrete masonry walls, the fire hose test is usually made on the original specimen after the full fire endurance test—and the wall usually passes this more severe test. Any distress caused by either the hose stream or double-load test is carefully noted and the rating adjusted or qualified.

Some confusion exists among architects, engineers, and builders as to the role of various organizations in the testing of assemblies and the recommendation of fire ratings. The following discussion of fire-testing and fire-rating agencies should be of help.

The function of Underwriters' Laboratories, Inc., is to examine and test materials and assemblies in order to determine if they comply with applicable safety standards. A fire-resistant assembly is tested in accordance with ASTM E119. As a result of this test, UL may certify that an assembly meets the stipulated specifications within close tolerances for a specific fire-resistance rating. Some of the individual products used in the assembly become eligible for one of the

UL followup inspection services. In the concrete field the UL certification service is usually limited to concrete masonry units and prestressed concrete elements. UL does not base its opinion on tests by other laboratories but may use such test data to supplement its own findings.

Among other laboratories equipped to conduct fire tests are those of the Portland Cement Association, National Research Council of Canada, National Bureau of Standards, Ohio State University, and Fire Prevention Research Institute.

The American Insurance Association, formerly the National Board of Fire Underwriters, does not conduct fire tests. AIA analyzes fire test data from all sources and recommends fire-resistance ratings. The National Building Code (United States) is the recommendation of the AIA.

The National Fire Protection Association compiles fire studies from all sources and publishes generally accepted ratings together with a vast amount of valuable fire protection information in its *Fire Protection Handbook.**

*See Ref. 69.

Fig. 3-26. A concrete masonry wall undergoing a fire test. A gas-fired furnace is on the opposite side of the wall. Jacks under the wall impose a working load on the wall during the test, and every phase of the test is closely observed and recorded by engineers.

Fig. 3-27. Fire hose stream test being applied at incandescent face of concrete masonry wall immediately after it is removed from the furnace.

Equivalent Solid Thickness

It has become nearly general practice in building codes to state the fire resistances required of concrete masonry walls in terms of the equivalent solid thickness. The transmission of heat to the unexposed surface of a wall is related to the thickness of solid material along the heat flow path. This relationship can be expressed by the following equation:*

$$R = (CV)^n$$

where R = the fire resistance as determined by heat transmission in hours or minutes

C = a constant that varies with the type of aggregate, design of the wall, and the units for expressing R and V

V = volume of solid material in the heat path per unit area of wall surface

n = an exponent, generally taken at 1.7

In the above equation V is in effect the average thickness of solid material. This has given rise to the term "equivalent solid thickness" or to the shorter version "equivalent thickness" (ET). The equivalent thickness is determined in various ways:

1. The Underwriters' Laboratories published a procedure in 1958 that employs an immersion tank. The unit to be tested is soaked in water for 24 hours. It is then removed, allowed to drain on a screen rack for one minute, and sponged with a cloth; after two minutes it is again immersed. The volume of water displaced is divided by the face area of the unit to determine the ET.
2. ET can also be determined from measurements. The thickness of a solid unit is the ET. For hollow units, the block machine manufacturer's drawing can be used, taking into account the taper of the cores.
3. Another method is to fill the cores and end voids with No. 10 lead shot (diameter of 0.07 in.) and then compare the volume of the shot to the volume of the block.**

The generally accepted values in the United States for *minimum* equivalent thicknesses for a wide range of fire resistance requirements are shown in Table 3-17. These conservative values apply only to wall construction in which framed-in structural members, if any, are of noncombustible materials. When inserted framing members are combustible, the equivalent solid thickness protecting such members must not be less than 93% (100% in Canada) of the total equivalent solid thickness of the wall in question. This applies to a wall supporting one or a series of wood beams or joists.

Plots of the values in Table 3-17, using logarithmic scales, are shown in Fig. 3-28. With approval of the code authority having jurisdiction, the relationships in Fig. 3-28 can be used in determining intermediate fire ratings and estimating increases in fire resistance due

Table 3-17. Equivalent Thickness and Fire Endurance of Concrete Masonry Walls (*United States*)*

Type of coarse aggregate in units	Minimum equivalent thickness, in., for fire-resistance rating indicated			
	1 hr.	2 hr.	3 hr.	4 hr.
Expanded slag or pumice	2.1	3.2	4.0	4.7
Expanded clay, shale, or slate	2.6	3.8	4.8	5.7
Limestone, scoria cinders, or air-cooled slag	2.7	4.0	5.0	5.9
Calcareous gravel	2.8	4.2	5.3	6.2
Siliceous gravel	3.0	4.5	5.7	6.7

*"Estimated ratings" from *Fire Resistance Ratings* published by the American Insurance Association (formerly the National Board of Fire Underwriters), March 1970, page 115 or E-100-1.

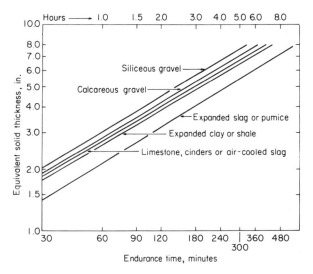

Fig. 3-28. Equivalent thickness and fire endurance of concrete masonry walls (United States). These are estimated ratings plotted from the values in Table 3-17.

to added equivalent thicknesses from fill materials, plaster coatings, and other composite layers.

Other values of equivalent thickness for specific fire ratings are available in the Uniform Building Code (UBC), BOCA Basic Building Code (Building Officials Conference of America), Southern Standard Building Code (SSBC), and the National Building Code of Canada (NBCC).† According to the UBC, the equivalent thickness may include the thickness of portland

---

*See Ref. 61.
**See details of method in Ref. 68.
†Also see Refs. 38, 39, 47, and 48.

Table 3-18. Fire Test Data of Concrete Masonry Walls Filled with Widely Different Materials*

Description of core fill				Fire endurance, min.
Type	Grading	Fineness modulus	Dry-rodded unit weight, pcf	
None	—	—	—	153
Calcareous sand	0—No. 4	2.68	114	383
Expanded shale	0—No. 4	2.27	68	382
Granular slag	0—No. 4	2.60	57	404
Diatomaceous earth	—	—	13.8	388

*Adapted from Table 15, Ref. 36. Core area is 45% of gross area and equivalent thickness is 4.2 in.

cement plaster or 1-1/2 times the thickness of gypsum plaster if applied according to that code. Equivalent thickness according to the NBCC is discussed later (under "Canadian Practice").

As stated previously, fire ratings are also determined by actual tests. Among the principal agencies for this work, Underwriters' Laboratories issues individual fire test reports and publishes a standard, Concrete Masonry Units (UL618).* This standard represents a compilation of actual fire tests, tabulated by minimum equivalent thickness.

Effect of Fill Material

Standard fire tests have shown that by filling the core spaces in hollow masonry units or the air space in cavity walls with dry granular materials, substantial reductions in the rate of heat transfer and increases in the fire endurance are realized. In many cases the fire endurance time may be more than doubled.**

For instance, hollow concrete block walls rated by Underwriters' Laboratories, Inc., as 2- and 3-hour fire-retardants may be upgraded to a 4-hour rating (walls having actual fire endurances of more than four but less than eight hours) by filling the core spaces with loose, dry expanded slag, clay, or shale, or water-repellent vermiculite masonry fill insulation. In fact, fire tests performed by the Portland Cement Association on a series of walls constructed of three-core 8×8×16 (actual dimensions) masonry units of calcareous sand and gravel concrete indicate that a fairly wide range of granular material may be used for core fill to increase fire resistance. Some details are given in Table 3-18.

In reinforced concrete masonry wall construction, portland cement grout is used to bond the steel reinforcing bars integrally with the masonry. In some instances grouting may be limited to those core spaces that contain steel. When grouting is so limited, the fire

endurance rating must be based on the performance of an ungrouted wall. Undoubtedly, the transmission of heat will be slowed through those wall areas containing grout, but the more rapid rise in the unexposed surface temperature between grouted areas will determine the fire endurance time.

There is some evidence that complete grout filling increases the fire resistance in about the same ratio that expanded slag, shale, perlite, or vermiculite core fill does. In computing the equivalent thickness, the Uniform Building Code allows the grout fill to be considered as making the block wall solid if all the cores are filled.

With respect to the transmission of heat from fire exposure, the grout in hollow unit masonry may be considered as one layer of a multilayered composite wall assembly. The fire resistance of composite wall construction may be approximated from the following equation:†

$$R = (\sqrt[n]{R_1} + \sqrt[n]{R_2})^n$$

where R = fire resistance (minutes) of the composite wall
R_1 = fire resistance (minutes) of the ungrouted wall
R_2 = fire resistance (minutes) of the grout based on its equivalent thickness and the type of aggregate used
n = an exponent, in this case 1.7

For example, assume that a wall is composed of 8×8×16-in. modular, expanded shale concrete units in which the cores are 45% of the gross volume. The equivalent solid thickness of the wall is therefore 7.625

---

*See Ref. 66.
**Canadian practice allows such filled walls; however, the fire endurance time may not exceed that of a wall of solid units of the same concrete type.
†See Ref. 61.

Table 3-19. Effect of Normal-Weight and Lightweight Aggregates on Fire Endurance*

Aggregate		Cement-aggregate proportions by volume**	Average weight of air-dry unit, lb.	Fire endurance, min.
Type	Fineness modulus			
All calcareous	4.50	1:7.7	49.9	141
50-50 mixture	4.25	1:7.4	37.5	157
All expanded shale	4.00	1:7.1	26.4	177

*Adapted from Table 7, Ref. 36.
**Based on dry-rodded volume of aggregate.

in. X (1.00 − 0.45) = 4.2 in. Assume also that the wall cores have been solidly filled with siliceous aggregate grout. The equivalent thickness of the grout, based on the core volume given, is 7.625 X 0.45 = 3.4 in. From Fig. 3-28 it will be found that the fire resistance of the ungrouted wall is 140 minutes and the fire resistance of the grout layer is 73 minutes. Substituting in the above equation, we find:

$$R = (\sqrt[1.7]{140} + \sqrt[1.7]{73})^{1.7} = 325 \text{ min.}$$

$$= 5 \text{ hr., } 25 \text{ min.}$$

If the wall above was constructed of siliceous aggregate concrete masonry units and siliceous aggregate grout, the ET would be 7.62 in., and the fire resistance obtained directly from Fig. 3-28 would be 300 minutes or 5 hours.

Effect of Sand Substitutions for Lightweight Fines

For various technical and economic reasons, producers of lightweight-aggregate concrete masonry units may occasionally find it desirable to substitute from 10 to possibly 40% (by volume) of the lightweight fines with natural sand. Generally, reduced fire resistance results from this practice due to an increase in the rate of heat transmission.

The actual effect is usually determined by a standard fire test. Some test data indicate that the reduction in fire resistance is directly related to the dry-rodded volumetric proportions of lightweight and normal-weight aggregates used in the concrete mixes. Table 3-19 shows some comparative results of fire tests conducted by PCA on walls of masonry units made with 100% calcareous aggregate, 100% expanded shale lightweight aggregate, and a 50-50 mixture of these aggregates. The fire endurance of the wall with mixed aggregates approximated the arithmetical average of

the fire endurances of the companion normal-weight and lightweight walls.

Canadian Practice

The "equivalent thickness" concept for estimating required fire resistance in concrete masonry construction has also been adopted in Supplement No. 2 of the National Building Code of Canada, as shown in Table 3-20. The ratings are slightly lower than those accepted in the United States.

The table footnote descriptions of the aggregates in Types S and N concretes apply to the coarse aggregates only. Coarse aggregate for this purpose means aggregate retained on a No. 4 sieve (0.187-in. openings). The use of siliceous sand as fine aggregate is not considered to affect the fire-resistance property of these concretes. If the nature of an aggregate cannot be determined accurately enough to place it in one of two groups, the aggregate should be considered as being in the group that requires a greater thickness of concrete for the same fire resistance.

The National Building Code of Canada also contains detailed provisions for the contributions of wall finishes to the fire resistance of a concrete masonry wall. Plaster or wallboard finishes do add to the fire resistance of walls if they are applied on the fire-exposed side. If they are applied only to the non-fire-exposed side of walls constructed of concrete masonry, the NBCC permits no increase in fire resistance over the same wall without plaster or wallboard because structural collapse may occur before the limiting temperature is reached. If plaster or wallboard is applied on both sides of a wall, the fire endurance of the wall assembly is limited to twice the fire endurance of the unfinished wall; structural collapse may occur before the limiting temperature is reached on the surface of the non-fire-exposed plaster or wallboard.

Tables 3-21 through 3-23 have been adapted from NBCC tables on the effect of wall finishes, and the

Table 3-20. Equivalent Thickness and Fire Endurance of Concrete Masonry Walls (*Canada*)*

Type of concrete	Minimum equivalent thickness, in., for fire-resistance rating indicated						
	1/2 hr.	3/4 hr.	1 hr.	1-1/2 hr.	2 hr.	3 hr.	4 hr.
S or N**	1.7	2.3	2.9	3.7	4.4	5.6	6.6
L₁20S†	1.6	2.1	2.6	3.4	4.0	5.1	6.0
L₁††	1.6	2.1	2.5	3.2	3.8	4.8	5.6
L₂20S†	1.6	2.1	2.5	3.2	3.7	4.6	5.3
L₂‡	1.6	2.1	2.5	3.1	3.6	4.4	5.1

*Adapted from Table 2.1.A of Supplement No. 2, 1975, National Building Code of Canada (Ref. 68), for both load-bearing and non-load-bearing concrete masonry walls.

**Type S concrete is that type in which the coarse aggregate is granite, quartzite, siliceous gravel, or other dense materials containing at least 30% of quartz, chert, or flint. In Type N concrete, the coarse aggregate is cinders, broken brick, blast-furnace slag, limestone, calcareous gravel, traprock, sandstone, or similar dense material containing not more than 30% of quartz, chert, or flint. Hollow concrete masonry units made with Type S or N concrete must have a minimum 28-day compressive strength of 1,000 psi.

†In Types L₁20S and L₂20S concretes, the fine portion of the lightweight aggregate is replaced by natural sand in a quantity not exceeding 20% of the total volume of all aggregates in the mix.

††In Type L₁ concrete, both fine and coarse aggregates are expanded shale.

‡In Type L₂ concrete, both fine and coarse aggregates are expanded slag, expanded clay, or pumice.

Table 3-21. Fire Endurance Time Assigned to Wallboard Membranes*

Description of finish	Time assigned to membrane, min.
1/2-in. fiberboard	5
3/8-in. Douglas fir plywood phenolic-bonded	5
1/2-in. Douglas fir plywood phenolic-bonded	10
5/8-in. Douglas fir plywood phenolic-bonded	15
3/8-in. gypsum wallboard	10
1/2-in. gypsum wallboard	15
5/8-in. gypsum wallboard	30
Double 3/8-in. gypsum wallboard	25
1/2 + 3/8-in. gypsum wallboard	35
Double 1/2-in. gypsum wallboard	40
Double 1/2-in. gypsum wallboard	50**
3/16-in. asbestos-cement + 3/8-in. gypsum wallboard	40†
3/16-in. asbestos-cement + 1/2-in. gypsum wallboard	50†
Composite 1/8-in. asbestos-cement, 7/16-in. fiberboard	20

*Adapted from Table 2.3.A of Ref. 68.

**16-gage, 1-in. square wire mesh must be fastened between the two sheets of wallboard.

†Value shown applies to walls only.

Table 3-22. Fire Endurance Time Assigned to Plaster Membranes*

Supporting material	Plaster thickness, in.	Time assigned to membrane, min., based on finish material				
		Portland cement-sand or portland cement-lime-sand	Portland cement-sand-asbestos fiber (3-lb./bag cement)	Gypsum-sand	Gypsum-wood fiber	Gypsum-perlite or gypsum-vermiculite
Wood lath	1/2	5	10	20	20	—
1/2-in. fiberboard	1/2	—	—	20	20	—
3/8-in. gypsum lath	1/2	—	—	35	35	55
3/8-in. gypsum lath	5/8	—	—	40	40	65
3/8-in. gypsum lath	3/4	—	—	50	50	80**
Metal lath	3/4	20	35	50	50	80**
Metal lath	7/8	25	40	60	65	80**
Metal lath	1	30	50	80	80	80**

*Adapted from Table 2.3.B of Ref. 68. Plaster and supporting material must be in accordance with CSA A82.30, Interior Fuiring, Lathing and Gypsum Plastering.
**Value has been limited to 80 minutes because the fire-resistance rating derived from these tables must not exceed 1-1/2 hours.

procedure for their use may be determined from the following schedule:

	Table for fire-exposed side	Table for non-fire-exposed side
Wallboard	3-21	3-23
Plaster applied on lath	3-22	3-23
Plaster applied directly	3-23	3-23

In Tables 3-21 and 3-22, the increased fire resistance is given directly in minutes. Table 3-23 gives multiplying factors to correct the thickness of the plaster or gypsum wallboard; the corrected thickness is then added to the equivalent thickness of the concrete masonry to determine the total fire rating of the wall.

As an example, assume hollow concrete masonry with Type L_2 concrete has an equivalent thickness of 2.60 in. and portland cement plaster 1 in. thick has been applied with metal lath to both sides. The plaster thickness on the non-fire-exposed side (according to Table 3-23) is multiplied by a factor of 1/2 to equal 0.50 in. Then adding this to the ET of the block gives an ET of 3.1 in., which (according to the last line of Table 3-20) equals a 1-1/2-hour fire rating. Furthermore, according to Table 3-22, the 1-in. plaster on metal lath for the fire-exposed side has a 30-minute endurance time and therefore the total wall assembly is given a 2-hour rating.

With plaster or wallboard applied only on the fire-exposed side of a wall, the contribution of the finish is limited by its ability to stay in place. If a plaster thickness exceeds 1-1/2 in., 16-gage wire mesh at 2 in.

Table 3-23. Multiplying Factor for Fire Resistance of Plaster Finish on Concrete Masonry*

Type of finish	Multiplying factor based on type of concrete**			
	Type N or S	Type $L_1$20S	Type L_1 or $L_2$20S	Type L_2
Portland cement-sand plaster	1	3/4	3/4	1/2
Gypsum-sand plaster or gypsum wallboard	1-1/4	1	1	1
Gypsum-vermiculite or gypsum-perlite plaster	1-3/4	1-1/2	1-1/4	1-1/4

*Adapted from Table 1.6.A of Ref. 68. Plaster must be in accordance with CSA A82.30-1965, Specification for Gypsum Plastering, Interior Furring and Interior Lathing.
**See Table 3-20 for key to types of concretes.

on center must be placed midway between the inner and outer surfaces of the plaster. The methods of fastening wallboard or lath to masonry walls permitted by the NBCC* are as follows:

1. By self-tapping drywall screws (12 in. oc) penetrating 3/8 in. into resilient steel furring channels that run horizontally (24 in. oc). The channels are

*See Ref. 68.

secured to concrete masonry by 1/2-in. concrete nails (16 in. oc).

2. By self-tapping drywall screws (12 in. oc) penetrating 3/8 in. into steel studs (not more than 24 in. oc for single-layer wallboard or not more than 16 in. oc for double-layer wallboard). The studs are secured by setting them into matching floor and ceiling runner channels adjacent to the unit masonry surfaces.

3. By lath nails (12 in. oc) penetrating 3/4 in. into nominal 1x2-in. wood strapping that is secured to the concrete masonry by 2-in. concrete nails (16 in. oc).

It should be noted that Method 3 is permitted for a noncombustible assembly by NBCC but not by some other codes.

Another successful method of attaching wallboard to block walls is by "glueing" or "laminating." However, although proven an acceptable attachment procedure in a National Research Council of Canada study,* it was not included in the NBCC because a generic specification for acceptable glues was not available. The procedure is as follows: (1) Apply a 3/8-in. bead of panel adhesive around the perimeter of the wallboard and across the diagonals. (2) After the wallboard is laminated to the block surface, secure it with concrete nails (approximately one nail per 2 sq.ft. of wall area).

---

*See Ref. 21.

Chapter 4
DESIGN AND LAYOUT OF CONCRETE MASONRY WALLS

The design of a concrete masonry wall depends on its required appearance, economy, strength, insulation, and acoustics. The layout of the wall involves other important considerations, such as the internal arrangement of components, modular planning, and provisions for shrinkage cracking control and weather resistance. All deserve careful planning if the wall is to successfully serve its intended purpose.

Fig. 4-1. Customized concrete block arranged in a dominating pattern.

Types of Walls

Concrete masonry walls may be classified as solid, hollow, cavity, composite, veneered, reinforced, or grouted. These classifications sometimes overlap, but the basic terminology and bonding directions remain the same, as shown in Fig. 4-2.

Solid Masonry Walls

Solid masonry walls (Fig. 4-3) are built of solid masonry units with all joints completely filled with mortar or grout. Facing units are usually brick or other solid architectural units that are laid with full head and bed joints. Backup units consist of solid masonry units laid with full head and bed joints.

If units with flanged ends are used, the end cavity must be filled with grout. The collar joints in exterior walls are completely filled by grouting, slushing, or backparging either the facing or backup units and then shoving them into place.

Structural bond between wythes is ensured by masonry headers, unit metal ties, continuous metal ties, or grout. Typical codes require that not less than 4% of the area of each wall face be composed of headers.

Headers usually consist of facing units laid transversely so they extend 3 to 4 in. into the backing. If the wall does not have headers extending completely through it, headers from the opposite sides overlap 3 to 4 in. The allowable vertical or horizontal distance between adjacent headers varies from 24 to 36 in., depending on the local code. Fig. 4-3 shows a solid wall in which the courses of each wythe overlap to create headers.

Hollow Masonry Walls

These walls (Fig. 4-4) are built of hollow or combined hollow and solid masonry units laid in face-shell mortar bedding. All horizontal and vertical edges of the face shells are mortared together.

Hollow masonry walls may be built in any required thickness with single or multiple wythes. Multi-wythe hollow masonry walls usually consist of two wythes—facing and backup—and may also be classified as composite walls. Bond between wythes is ensured by masonry headers, unit metal ties, continuous metal ties, or grout. All collar joints are filled with mortar.

Cavity Walls

A cavity wall (Fig. 4-5) consists of two walls separated by a continuous air space 2 to 3 in. wide and tied together by rigid metal ties embedded in the mortar joints of both walls. The facing wall usually consists of

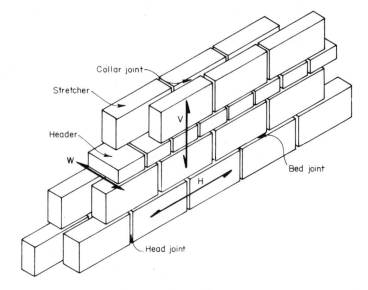

Fig. 4-2. Basic terms and bonding directions.

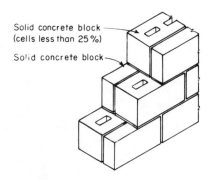

Fig. 4-3. Solid masonry wall.

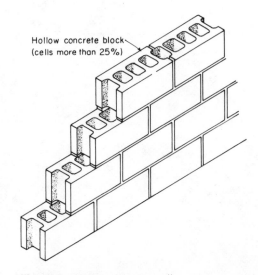

Fig. 4-4. Hollow masonry wall.

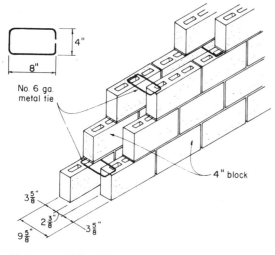

No. 6 ga. metal tie

4" block

(a) 10-in. wall of 4-in. block

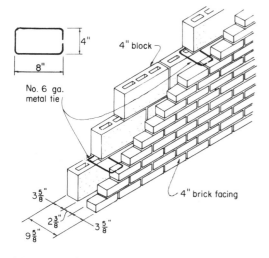

4" block

No. 6 ga. metal tie

4" brick facing

(b) 10-in. wall of block and 4-in. brick

Fig. 4-5. Cavity walls.

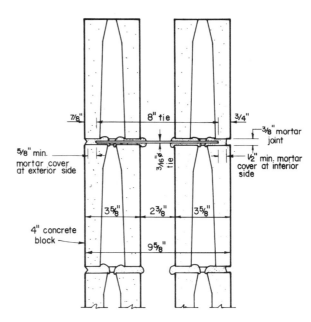

(c) Detailing of metal tie in 10-in. wall

one wythe of solid or hollow masonry units 3-1/2 to 4 in. thick. The backing may be a single- or multi-wythe solid or hollow masonry wall. The thickness of the backing may be equal to or greater than that of the facing, depending on such structural requirements as wall height and the loads to be carried. Usually the cavity wall is designed so that all the vertical loads are carried by the backing; the outer wall serves as a weather-protective facing. Being tied together, both walls act to resist the wind, although not necessarily equally.

In areas of severe weather exposure, the wall cavity offers three main advantages:

1. It increases the insulating value of the wall and permits use of insulation within the wall.

2. It prohibits the passage of water or moisture across the wall.

3. It prevents the formation of condensation on interior surfaces; therefore, plaster may be applied directly to the masonry without furring or the interior surface may be used as the finished wall without plastering.

Insulation placed within the cavity consists of mats, rigid boards, or non-water-absorbent fill material such as water-repellent vermiculite or silicone-treated perlite. Mats or rigid boards may be glass fiber, foamed glass, or foamed plastics. A vapor barrier or damp-proofing is required on the cavity face of the inner wall unless waterproofed insulation is used or the insulating rigid boards are held at least 1 in. away from the exterior wall.

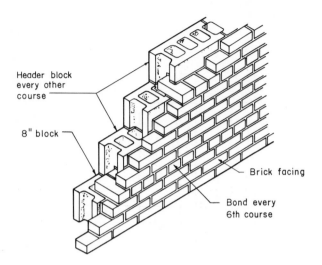

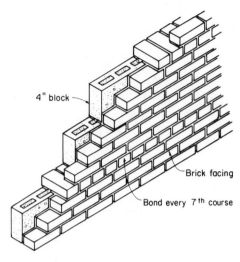

(a) 12-in. wall--block and brick, 6th-course bonding

(c) 8-in. wall--block and brick

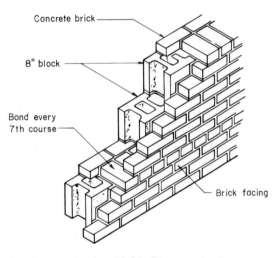

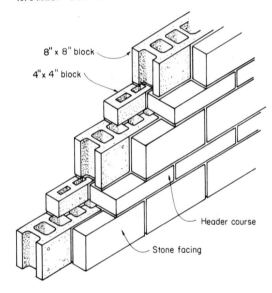

(b) 12-in. wall--block and brick, 7th-course bonding

Fig. 4-6. Composite walls.

(d) 12-in. wall--block and stone

Composite Walls

A composite wall (Fig. 4-6) is a multi-wythe wall having at least one of the wythes dissimilar to the other(s) with respect to type or grade of masonry unit or mortar, although each wythe contributes to the strength of the wall. Brick or stone may be bonded to the concrete masonry. Since a composite wall containing different materials has wythes that are *bonded* together, it is not considered a veneered wall.

The masonry headers consist of solid or hollow masonry units lapping over the courses below. In building 8- and 12-in. walls with brick facing and concrete block backup, every seventh course of brick is a header course (Figs. 4-6b and c). The 12-in. walls can also be bonded every sixth brick course by using concrete block header units (Fig. 4-6a).

Veneered Walls

It is common practice in residential construction to use masonry veneer as a non-load-bearing siding or facing material over a wood frame, as detailed in the Appendix (Fig. A-43, page 203). Designed to carry its own weight only, veneer is anchored but not bonded to the backing.

The purpose of the veneer is to provide a durable, attractive exterior finish that will prevent entrance of water or moisture. An air space of at least 1 in. is provided between the veneer and the wood backing to give additional insurance against moisture penetration and heat loss. Flashing and weepholes are provided at the bottom of the air space to eliminate water that may penetrate the veneer.

Metal ties anchoring the veneer to the backing are

Fig. 4-7. A reinforced concrete masonry wall veneered with brick. The collar joint is not filled with mortar.

usually 22-gage corrugated, galvanized steel strips 7/8 in. wide. Building code requirements for spacing of such ties vary widely, but an average value would be 16 in. vertically and 32 in. horizontally. During a few tests on corrugated metal ties,* the load at failure was on the order of 60 lb. per tie for compression, the governing case.

Veneer may also be anchored to the backing by grouting it to paperbacked, welded-wire fabric attached directly to the wood studding. The thickness of grout between the backing and the veneer is at least 1 in. No sheathing is required, although it may be added for stiffness, and the need for flashing and weepholes at the base of the wall is eliminated. This type of construction is commonly called reinforced masonry veneer. Another type of veneered wall with reinforcement appears in Fig. 4-7.

Reinforced Masonry Walls

Reinforced concrete masonry walls (Fig. 4-8) are used in cases of high stress concentrations, or in areas of high winds or earthquake probabilities. Embedment of steel in grouted vertical and horizontal cavities gives the wall increased strength. This permits the use of higher design stresses and an increase in the distance between lateral supports. Single or multiple

wythes may be used.

Single-wythe walls consist of hollow masonry units laid in face-shell mortar bedding. The vertical cores are aligned to form continuous, unobstructed vertical spaces for the reinforcement to be placed and grouted. Two-core block are preferred to three-core block because of the ease in placing reinforcement and grout.

Multi-wythe walls usually consist of two wythes. In two-wythe construction, the wythes are erected with a 1- to 6-in.-wide continuous air space between them, depending on structural requirements, and may have hollow or solid units. Exterior walls may consist of a solid facing and a backing of hollow units, while

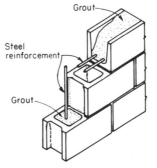
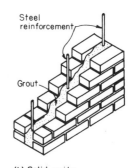

(a) Hollow units   (b) Solid units

Fig. 4-8. Reinforced walls.

*See Ref. 19.

75

interior walls may consist of two wythes of hollow units. Reinforcement is placed in the space between the wythes and grouted solid. Depending on site conditions, grouting may be performed intermittently as the wall is erected (low-lift grouting) or completely after erection (high-lift grouting).

Units used in three-wythe construction are usually solid and erected by the low-lift grouting method. The two outer wythes are erected first; reinforcement is placed in the cavity between the wythes; and grout is poured in the cavity. The units of the middle wythe are then "floated" (embedded) into the grout so that 3/4 in. of grout surrounds the sides and ends of each unit.

In high-lift grouted construction the grout space is not less than 2 or 3 in. wide, depending on the code. Wythes are tied with No. 9 rectangular ties 4 in. wide and 2 in. narrower than the nominal wall thickness. Codes do not permit the use of kinked or crimped ties in these walls. Spacing of ties is usually 24 in. horizontally and 16 in. vertically for a running bond pattern or 12 in. vertically for a stacked bond pattern. Wall ties are not necessary in low-lift grouted construction.

Usually vertical and horizontal reinforcement consists of bars of 3/8-in. minimum diameter. Some building codes permit the use of prefabricated-type continuous joint reinforcement as part of the horizontal steel. However, most building codes require that the minimum area of reinforcement in either direction be not less than 0.07% of the gross cross-sectional area of the wall and that the sum of the percentages of horizontal and vertical reinforcement be at least 0.2%.

Special-shape units (Fig. 4-9) have been developed for use in single-wythe reinforced masonry construction. The most common one is the open-end unit shown in Fig. 4-9a. The advantage of this unit is that it can be laid around vertical steel rather than threaded down over the rods. In bond-beam units (Fig. 4-9b and c), low webs not only permit grout to flow both horizontally and vertically, but also facilitate placement of horizontal steel. Some bond-beam units are manufactured with a closed bottom for use in lintel construction; some have depressions in the cross webs to maintain the appropriate steel spacing; and some may also be open-end types (Fig. 4-9b). To permit easy removal of part of the face shell for cleanout openings at the wall base, some units are scored.

Units used in reinforced concrete masonry sometimes have higher compressive strength than normal load-bearing units; the compressive strength generally ranges from 3,000 to 5,000 psi on the net area. Normally, block producers only stock units that have strengths conforming to ASTM specifications. The local market should be studied for availability of high-strength units.

Grouted Masonry Walls

Grouted masonry walls are similar to reinforced masonry walls but do not contain reinforcement. Grout is sometimes used in load-bearing wall construction to give added strength to hollow walls by filling a portion or all of the cores. It is also used in filling bond beams and, occasionally, the collar joint of a two-wythe wall.

Bonding of Plain Masonry

All of the concrete block or brick in a masonry wall must be bonded (held together) to form a continuous mass. Although several methods may be used, bonding is treated as follows in each of the three directions shown previously (Fig. 4-2):

1. *Horizontally in the plane of the wall* (H). The mortar in the bed joint exerts a shearing bond on the top and bottom of the stretchers, which overlap to transfer loads and stresses across the head joints. Also, mortar in the head joint tends to hold the wall together. In a stacked bond pattern, codes require the horizontal joints to be reinforced with the equivalent of a No. 9 gage wire for each 4-in. width of masonry unit at

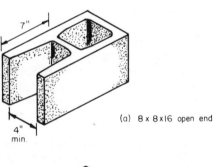

(a) 8 x 8 x16 open end

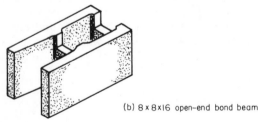

(b) 8 x 8x16 open-end bond beam

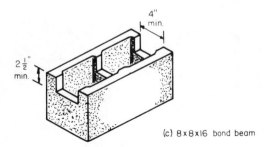

(c) 8 x 8x16 bond beam

Fig. 4-9. Special units for reinforced walls.

vertical spacings of 16 in. The same reinforcing rules are recommended where there is only a small overlap of one course over the other. The reinforcement is no longer required where 75% of the units in any vertical plane overlap the units below by 1-1/2 in. or half the height of the units, whichever is greater.

2. *Vertically in the plane of the wall* (V). Bonding is achieved by the tensile strength of the mortar and its adhesion to the masonry unit. The force of gravity assists this and helps offset tensile stresses caused by wind or other lateral forces. Because adhesive tensile strength of a mortar joint is only a small fraction of the compressive strength, careful engineering analysis of all tensile stresses is required, as discussed in Chapter 3.

3. *Horizontally across the width (thickness) of a two-wythe wall* (W). The tensile bond of mortar in a collar joint is not credited; custom and codes require more positive measures such as masonry unit headers or metal ties. On the other hand, low-lift grouted construction requires neither headers nor ties.

Masonry Headers

Masonry headers consist of stretcher units laid transversely to overlap units of the adjacent wythes or specially shaped header units. Usually they are placed in continuous courses.

Headers have certain disadvantages over metal ties. When headers are used in a bonded multi-wythe masonry wall having backup units of concrete masonry and a brick facing, care must be taken to maintain bond between the headers and the mortar. Immediately following the laying of the header course, heavy backup units are set on the brick headers, temporarily loading the headers off center (Fig. 4-10). The mortar below the headers should have time to stiffen or the headers may settle unevenly, causing a fine crack to develop at the external face between the headers and the facing underneath.

Furthermore, backup and facing units may have different thermal and shrinkage properties and consequently different volume changes. Excessive vertical shrinkage of the concrete masonry backup (compared to the facing) may load the headers eccentrically, adding to rupture of the bond between the headers and the mortar at the external face.

The advantage of metal ties over masonry headers is sufficient flexibility to accommodate the differential movements between adjacent wythes, thus relieving stresses and preventing cracking. They are recommended when resistance to rain penetration is important or where wide differences in the physical characteristics of facing and backup exist.

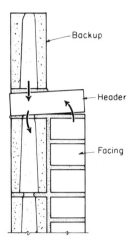

Fig. 4-10. Mortar bond can be impaired by the use of masonry headers.

Unit Metal Ties

These ties consist of corrosion-resistant rods or wire embedded in the horizontal mortar joints and engaging all wythes. They are usually made of galvanized steel, but they may be copper-coated or made of stainless steel.

In cavity walls the codes generally require that metal ties be 3/16-in.-diameter steel rods or metal wire ties of equivalent strength and stiffness. Ties made of No. 6 gage wire are essentially the same in diameter as the steel rods. According to a National Bureau of Standards test report,* the tensile or compressive strength of a 3/16-in.-diameter straight tie in a cavity wall exceeds 1,200 lb. when the mortar has a cube strength of 1,330 psi. A crimp (often called a "drip") in the tie reduces the strength about 50%.

Unit metal ties used with cavity walls very frequently have a crimp located in the center of the tie, as shown in Fig. 4-11a. Its function is to cause any water that finds its way into the cavity to drip off at the crimp before reaching the inner wythe. Although crimps reduce the buckling strength of ties, they have become almost a standard feature in cavity walls. In high-lift grouted reinforced walls the crimp is not allowed.

Fig. 4-11a also shows a few commercial tie sizes. Ties used with solid masonry are generally bent in a "Z" shape with 2-in.-long legs at 90 deg. Although some codes allow the Z shape for ties used with hollow masonry, they are usually bent to a rectangular shape 4 in. wide. Tie length is such that the ends are embedded in the face-shell mortar beds at the outside faces of the wall.

In cases where commercial tie sizes will not fit properly in the outer face-shell mortar beds, a mortar

---

*See Ref. 19.

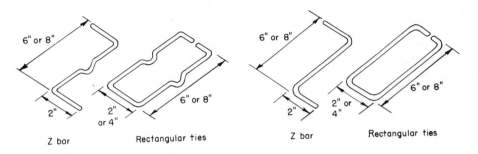

With crimp for cavity walls   For solid, composite, reinforced, or grouted walls

(a) Regular ties

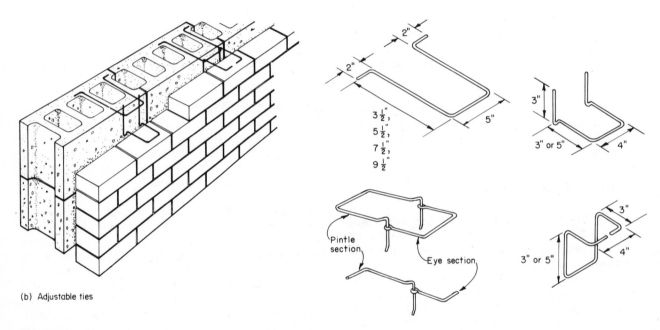

(b) Adjustable ties

Fig. 4-11. Unit metal ties.

bed base is provided by strips of metal lath laid in the joints receiving the ties; otherwise ties are placed over the head joints or over the webs. A cover of mortar at least 5/8 in. thick is required at any face exposed to the weather (Fig. 4-5c). The National Building Code of Canada requires that the tie in a hollow wall be placed on a web and completely embedded in mortar.

Ties are usually spaced not more than 3 ft. horizontally and 18 in. vertically, with each level staggered from the level above or below. To ensure adequate bonding of the wythes at openings or on both sides of control and isolation joints, unit ties are installed around the perimeter within 12 in. of the opening or joint.

Unit metal ties may be also of the adjustable type, as shown in Fig. 4-11b. These ties simplify the erection of multi-wythe walls by allowing the mason to erect the wythes independently instead of simultaneously, as with ordinary ties. They also permit adjustment for

differences in level between courses. The eye or loop section of the adjustable tie is installed in the first wythe erected. It is preferably located at a head joint or cross web in hollow masonry, with the eyelets or loop close to the face of the wythe. The pintle section, which is always installed in the second wythe, is inserted either up or down.

Continuous Metal Ties

Continuous metal ties are called prefabricated joint reinforcement, mesh or, more commonly, *joint reinforcement*. Consisting of two or more parallel longitudinal wires to which cross wires are welded (Fig. 4-12), joint reinforcement may be used for the following reasons:

1. To act as horizontal reinforcement, at a vertical spacing of 16 in., for units laid with stacked bond.

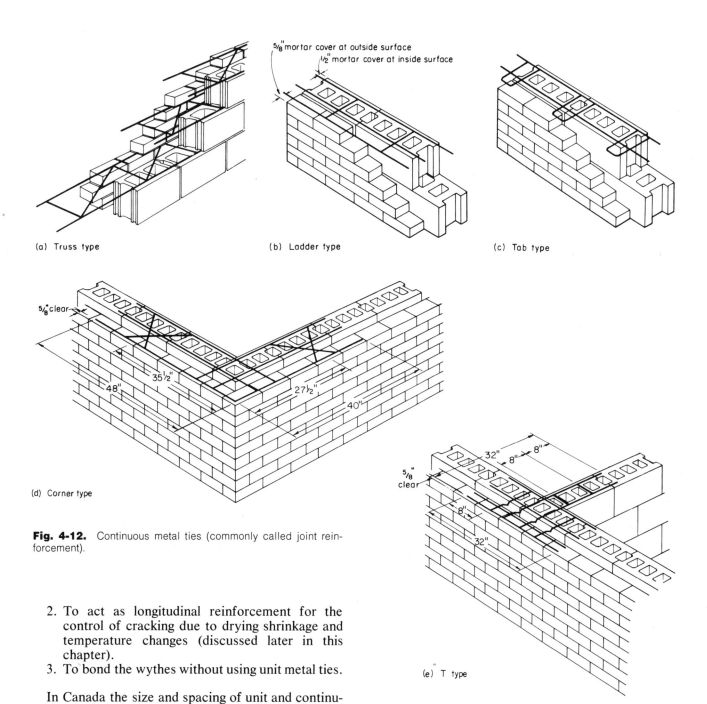

⅝" mortar cover at outside surface
½" mortar cover at inside surface

(a) Truss type

(b) Ladder type

(c) Tab type

⅝" clear

48"   35½"   27½"   40"

(d) Corner type

Fig. 4-12. Continuous metal ties (commonly called joint reinforcement).

⅝" clear   32"   8"   8"   8"   32"

(e) T type

2. To act as longitudinal reinforcement for the control of cracking due to drying shrinkage and temperature changes (discussed later in this chapter).
3. To bond the wythes without using unit metal ties.

In Canada the size and spacing of unit and continuous metal ties have the same requirements. However, ACI Committee 531* and others require one cross wire for each 2 sq.ft. of wall surface, with vertical spacing limited to 16 in.

Cross wires are corrosion-resistant and may be crimped for use with cavity walls. They are welded diagonally or perpendicularly to the longitudinal wires, usually at 16-in. spacings. The longitudinals are preferably deformed (rather than smooth) wire to obtain a

*See Ref. 52.

better bond with the mortar. Some manufacturers have developed continuous rectangular-tie assemblies (also called tab ties) that have rectangular ties welded at fixed intervals to longitudinal wires, as shown in Fig. 4-12c.

Continuous metal ties are available in any width, usually 2 in. narrower than the wall, and in commercial lengths of 10 to 12 ft. The longitudinal and cross wires are generally No. 8 or 9 gage; however, 3/16-in.-diameter wire assemblies are sometimes used. Usually No. 9 gage ties are used with 3/8-in. mortar joints; No. 8 gage, with 7/16-in. joints; and 3/16-in. wire, with 1/2-in. joints. A mortar cover of at least 5/8 in. is required at any face exposed to the weather.

Wall Patterns

Exposed concrete masonry is an attractive finished wall material for both exteriors and interiors of homes, churches, schools, and public and commercial buildings. One reason for the popularity of concrete masonry is the broad choice of sizes, shapes, textures, and colors. A variety of architectural effects may be obtained by: (1) varying the pattern in which units are laid and (2) applying different treatments to the mortar joints.

For example, if a long, low look is desired, 2-in.-high units 16 in. long will accentuate the horizontal lines. The opposite effect can be achieved by using other size units evenly placed one atop the other to emphasize the vertical lines (stacked bond pattern). Concrete block can also be laid in staggered (running bond), diagonal, and random patterns to produce almost any result the designer may be seeking.

Customized architectural concrete masonry units have enjoyed an immense popularity not only for use as individual profiles (Fig. 4-14), but to enhance or supersede the mortar pattern (Fig. 4-15). The designer can use his ingenuity to create any of a multitude of pattern arrangements and, by using block with a visible third dimension, an infinite diversity of effects.

In some wall treatments all the joints are accentuated by deep tooling; in others only the horizontal joints are accented. In the latter treatment the vertical joints are tooled, refilled with mortar, and then rubbed flush (after the mortar has partially hardened) to give the joints a texture similar to that of the concrete masonry units. This treatment makes the horizontal joints stand out in relief. It is well suited to walls where strong horizontal lines are desired. If an especially massive effect is sought, every second or third horizontal or vertical joint can be accented by having all other joints, both horizontal or vertical, refilled with mortar (after tooling) and rubbed flush.

Numerous wall bond patterns are illustrated in Fig. 4-16. Variations of these patterns may be created by projecting or depressing the faces of some units from

Fig. 4-13. Roman concrete brick (*left*) and diagonal stacking of 8×8-in. block (*right*).

the overall surface of the wall—or by substituting screen block, split block, or customized architectural units with three-dimensional faces, as discussed in Chapter 1 (page 15).

Fig. 4-14. The fleur-de-lis in concrete masonry enhance a garden setting.

Fig. 4-15. Customized sculptured block create a rhythmic pattern that overshadows the joint pattern.

(a) Running bond, 8×16 in.

(d) Running bond, 4×16 in.

(b) Offset bond, 8×16 in.

(e) Offset bond, 4×16 in.

(c) Offset bond, 8×16 in.

(f) Coursed ashlar, 4×16, 8×16 in.

Fig. 4-16. Forty-two patterns for concrete masonry walls. Some of the patterns illustrated require units of a size or shape that may not be produced in all areas. Local concrete masonry producers should be consulted as to available units.

(g) Coursed ashlar, 4×16, 8×16 in.

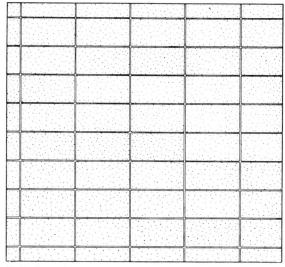

(j) Horizontal stacking, 8×16 in.

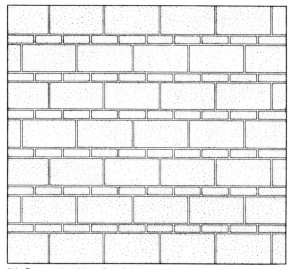

(h) Coursed ashlar, 8×16 in., brick size.

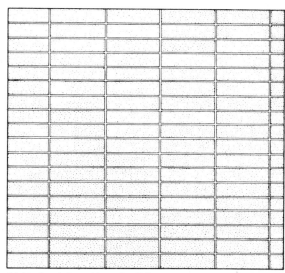

(k) Horizontal stacking, 4×16 in.

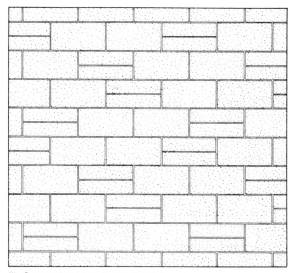

(i) Coursed ashlar, 4×16, 8×16 in.

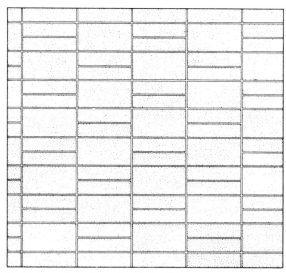

(l) Horizontal stacking, 4×16, 8×16 in.

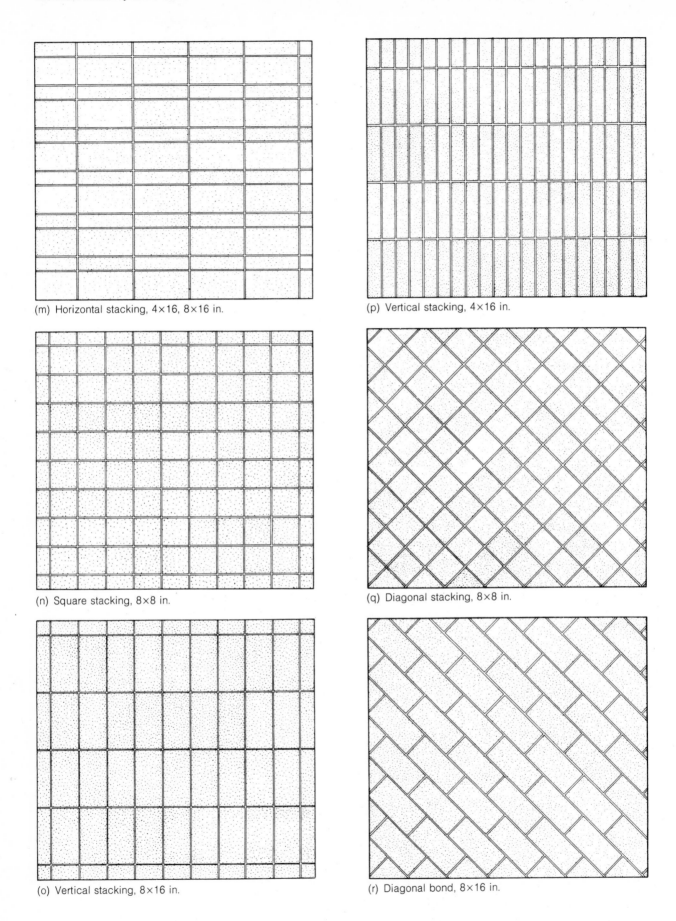

(m) Horizontal stacking, 4×16, 8×16 in.

(p) Vertical stacking, 4×16 in.

(n) Square stacking, 8×8 in.

(q) Diagonal stacking, 8×8 in.

(o) Vertical stacking, 8×16 in.

(r) Diagonal bond, 8×16 in.

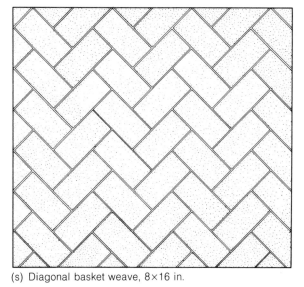

(s) Diagonal basket weave, 8×16 in.

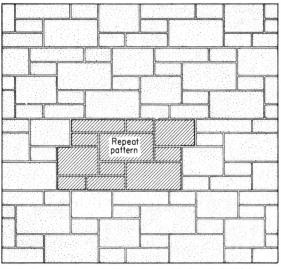

(v) Patterned ashlar, 4×8, 4×12, 4×16, 8×12, 8×16 in.

(t) Diagonal basket weave, 8×16 in.

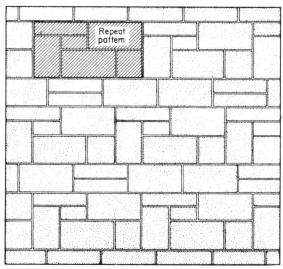

(w) Patterned ashlar, 4×8, 4×16, 8×8, 8×12, 8×16 in.

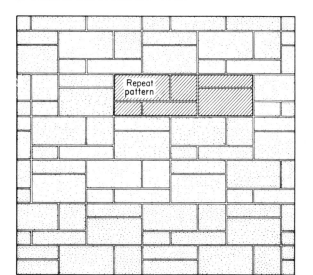

(u) Patterned ashlar, 4×8, 4×16, 8×8, 8×16 in.

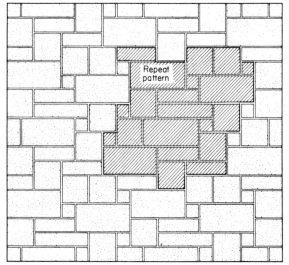

(x) Patterned ashlar, 4×4, 4×12, 4×16, 8×8, 8×16 in.

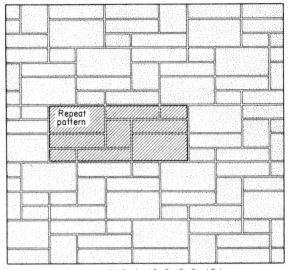

(y) Patterned ashlar, 4×8, 4×16, 8×8, 8×16 in.

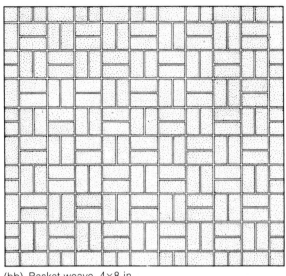

(bb) Basket weave, 4×8 in.

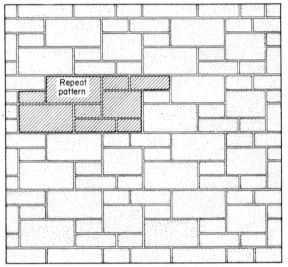

(z) Patterned ashlar, 4×8, 4×12, 8×12, 8×16 in.

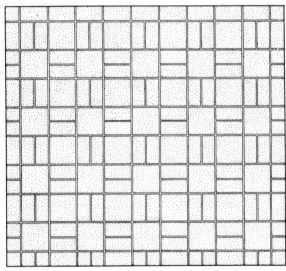

(cc) Basket weave, 4×8, 8×8 in.

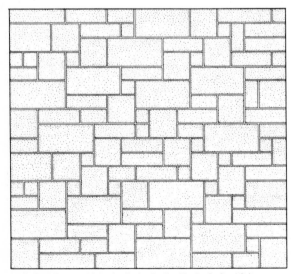

(aa) Random ashlar, 4×4, 4×8, 4×12, 8×8, 8×16 in.

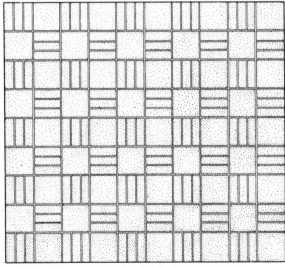

(dd) Basket weave, 8×8 in., brick size.

(ee) Basket weave, brick size.

(hh) Basket weave, 4×8, 8×16 in.

(ff) Basket weave, 4×8, 4×16 in.

(ii) Basket weave, 4×16 in.

(gg) Basket weave, 4×8, 8×16 in.

(jj) Basket weave, 8×8, 8×16 in.

(kk) Basket weave, 8×16 in.

(nn) Basket weave, 4×4, 8×16 in.

(ll) Basket weave, 8×16 in.

(oo) Basket weave, 8×8, 8×16 in.

(mm) Basket weave, 4×8, 4×16, 8×16 in.

(pp) Basket weave, 8×8, 8×16 in.

Jointing and Control of Cracking

It is well known that various building materials are subject to movement. The movement of concrete masonry walls is due to changes in temperature, changes in moisture content, contraction due to carbonation, and movements of other parts of the structure. When concrete masonry units are bonded together by mortar to form a wall, any restraint that will prevent the wall from expanding or contracting freely will set up stresses within the wall.

Restraint against expansion generally results in low stresses in relation to the strength of the material and rarely causes damage to concrete masonry walls. Moreover, expansion of the walls is offset by shrinkage from carbonation and drying of the units. Expansion joints are not necessary in concrete masonry except where required for the length or configuration of the building.* It should be noted, however, that concrete masonry used as backup for a clay brick facing does require occasional expansion joints since clay brick walls need them.

When contraction of the concrete masonry units is prevented, tensile stresses gradually build up within the wall. If these stresses exceed the tensile strength of the unit, the bond strength between the mortar and the unit, or the shearing strength of the horizontal mortar joint, cracks will occur to relieve the stresses (Fig. 4-17). Cracks usually disfigure the wall and cannot be easily concealed. Also, they affect the lateral stability of adjacent wall sections and must be calked for a weather seal. They can be controlled, as discussed later.

Shrinkage Due to Moisture Loss

Of the principal causes of volume changes, shrinkage due to moisture loss is the greatest concern to the designer. The drying shrinkage of concrete masonry units is primarily affected by the type of aggregate used, the method of curing, and the method of storage. Units made with sand and gravel aggregate will normally show the least shrinkage, and those with pumice the highest. The difference between the moisture content of the masonry units during construction and after the building is occupied will determine the amount of shrinkage in the wall.

There is evidence that high-pressure steam curing (autoclaving) of concrete masonry units will reduce shrinkage to approximately one-half the shrinkage of low-pressure steam-cured units. In pumice units, however, the reduction in shrinkage will be approximately one-third.

Proper drying of the units before they are laid in the wall will reduce the potential shrinkage of the wall. The degree of dryness will vary according to locality

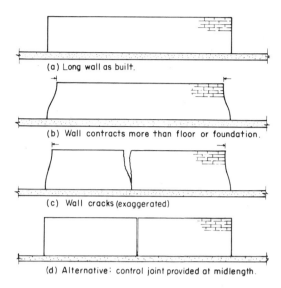

(a) Long wall as built.

(b) Wall contracts more than floor or foundation.

(c) Wall cracks (exaggerated)

(d) Alternative: control joint provided at midlength.

Fig. 4-17. Cracking of long wall by contraction (exaggerated).

and use. Ideally, the units should be laid with a moisture content corresponding to or preferably slightly below the average annual relative humidity of the outside air for the locality or for the ambient atmospheric condition to which they will be exposed.

As discussed in Chapter 1, ASTM and CSA standards classify concrete masonry units as moisture-controlled units or non-moisture-controlled units. Moisture-controlled units have been dried during production and then kept dry to a certain extent so their moisture content does not exceed the requirements spelled out in the standards. Non-moisture-controlled units, as the name implies, have not been subject to limitations on their moisture content during production. However, some codes or standards limit their moisture content at the time of installation to 40% of the total absorption or do not allow their use. In any case, since moisture content is the predominant factor affecting shrinkage, it is essential that concrete masonry units be kept dry until laid into the wall.

In common construction practice, two methods are used to accommodate shrinkage: minimizing the amount of stress buildup by means of discontinuities in the length of the wall (control joints), and minimizing the width of cracks by means of suitable restraints (joint reinforcement or bond beams). These two methods can be used separately or together, as will be explained.

Types of Control Joints

Control joints are continuous, vertically weakened sections built into the wall. If stresses or wall move-

_____
*See Ref. 46.

ments are sufficient to crack the wall, the cracks will occur at the control joints and thus be inconspicuous.

A control joint must permit ready movement of the wall in a longitudinal direction and be sealed against vision, sound, and perhaps weather. In addition, it may be required to stabilize the wall laterally across the joint by means of a shear key.

There are a number of types of control joints built into concrete masonry walls, but the most preferred types are the Michigan, the tongue-and-groove, and the premolded gasket. Fig. 4-18 shows the so-called Michigan type of control joint. It uses conventional flanged units. A strip of building paper is curled into the end core covering the end of the block on one side of the joint and, as the block on the other side of the joint is laid, the core is filled with mortar. The filling bonds to one block but the paper prevents bond to the block on the other side of the control joint. Thus, the control joint permits longitudinal movement of the wall while the mortar plug transmits transverse loads.

Figs. 4-19 and 4-20 show the tongue-and-groove type of control joint. The special units are manufactured in sets consisting of full- and half-length units. The tongue of one special unit fits into the groove of another special unit or into the open end of a regular flanged stretcher. The units are laid in mortar exactly

the same as any other masonry units, including mortar in the head joint; this is done so the mason can maintain bond more easily. Also, part of the mortar is allowed to remain in the vertical joint to form a backing against which the calking can be packed. The tongue-and-groove units provide excellent lateral stability for the wall.

Fig. 4-21 shows another type of control joint. It is made by installing a fairly stiff premolded rubber insert in the vertical joint.

Still another type of control joint is made with two jamb block, as shown in Fig. 4-22. This method, however, upsets the modular planning (discussed later), making layout more difficult. Lateral stability is

Fig. 4-19. Special units for tongue-and-groove type of control joint.

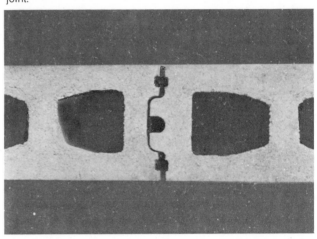

Fig. 4-20. Tongue-and-groove units for control joints are made in full- and half-length sizes.

Fig. 4-18. Michigan-type control joint.

attained by putting a "Z" bar across the joint as shown, and only the face shells are mortared.

In still another type of control joint, regular open-end units are tied with a Z bar across the joint (Fig. 4-23) or a 10- or 12-in. pencil rod across each face shell. When a pencil rod is used, the rod must be greased on one side of the joint to prevent bond.

All of these control joints are first laid up in mortar just the same as any other vertical mortar joint. However, if a control joint is to be exposed to view or the weather, the mortar should be permitted to become quite stiff before a recess is raked out of it to a depth of about 3/4 in. (Fig. 4-24). The mortar remaining in the control joint forms a backing to confine a calking compound or similar elastic weathertight material. First, however, to prevent absorption of oils from certain calking compounds, the side faces of the raked joint should be primed with shellac, aluminum paint, or other sealer, but the inner face of the joint should be greased or given some other bond-breaker. Then the calking is applied by using a calking gun or, for knife-grade compound, a pointing trowel (Fig. 4-25). Care must be taken not to smear the calking onto the face of the wall.

Fig. 4-23. Z bar creates shear strength at control joint.

Fig. 4-24. At a control joint, mortar is raked out to a depth of about 3/4 in. Either a pointing trowel (shown here) or a special wheeled rake may be used.

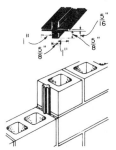

(a) Control joint at concrete pilaster or column

(b) Control joint used with standard sash block

Fig. 4-21. Premolded control joint insert provides lateral support.

Fig. 4-22. Control joint created with two jamb block.

Fig. 4-25. Calking applied to a control joint.

Location of Control Joints

Various rules for locating control joints have been developed from experience and will probably continue to be refined. Since there are many possible layouts of walls and partitions with their openings for doors, windows, and ducts, some judgment must be used in determining where the joints should be built.

ACI recommendations for spacing of control joints are given in Table 4-1. In earthquake regions the U.S. Army, Navy, and Air Force* limit the maximum spacing between control joints in reinforced masonry construction to 50 ft. and the distance of a joint from a corner to 25 ft.

Control joints should also be located at the following points of weakness or high stress concentrations:

1. At all abrupt changes in wall height.
2. At all changes in wall thickness, such as those at pipe or duct chases and those adjacent to columns or pilasters (Figs. 4-26 and 4-27).
3. Above joints in foundations and floors.
4. Below joints in roofs and floors that bear on the wall.
5. At a distance of not over one-half the allowable joint spacing from bonded intersections or corners.
6. At one or both sides of all door and window openings unless other crack control measures are used, such as joint reinforcement or bond beams.

All large openings in walls should be recognized as natural and desirable joint locations. Although some adjustment in the established joint pattern may be required, it is effective to use vertical sides of wall openings as part of the control joint layout. Under windows the joints usually are in line with the sides of the openings. Above doors and windows the joints must be offset to the end of the lintels. To permit movement, the bearing of at least one end of the lintel should be built to slide (Fig. 4-28). Plastic or bitumi-

*See Ref. 60.

nous sheets or other suitable material should be used for a slip plate.

Openings less than 6 ft. wide require a control joint along one side only, but openings of more than 6 ft. should have joints along both sides (Fig. 4-29). A control joint between two windows should be avoided since it will not function properly (Fig. 4-30).

To avoid the occurrence of cracks due to differential movement between concrete masonry and struc-

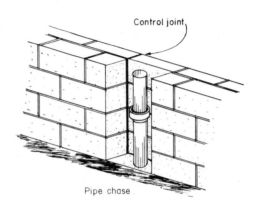

Fig. 4-26. A control joint should be located at a pipe chase or any other abrupt change in wall thickness.

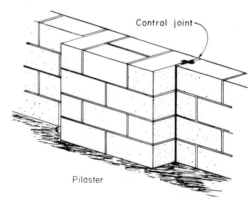

Fig. 4-27. A pilaster edge is a good location for a control joint.

Table 4-1. Maximum Spacing of Control Joints in Nonreinforced Masonry*

Maximum spacing of joint reinforcement,** in.	Maximum spacing of control joints	
	Panel length/height	Panel length, ft.
None	2	40
24	2.5	45
16	3	50
8	4	60

*Adapted from Table 20.3.2 of Ref. 52. The indicated spacings are for moisture-controlled units; if non-moisture-controlled units are used, the spacings should be reduced by one-half. Where the wall is solid-grouted, the spacings should be one-third less.

ACI Committee 531 recognized that by use of units of lower drying shrinkage and/or lower moisture content, and by consideration of additional factors such as tensile strength, extensibility, dead load, and mortar qualities, it may be possible to safely exceed these limits. However, the committee felt it did not have sufficient data available to formulate generalized recommendations that take these factors into consideration.

**Continuous metal ties with a minimum of two No. 9 gage longitudinal wires.

Fig. 4-28. Sliding bearing for a lintel.

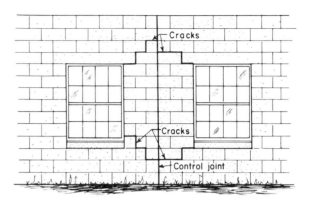

Fig. 4-30. The wrong place for a control joint because cracks may seek a path of less restraint.

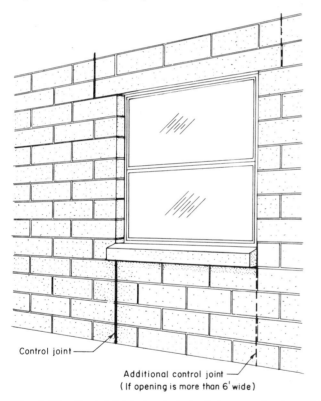

Fig. 4-29. Control joints located at window opening to avoid random cracking.

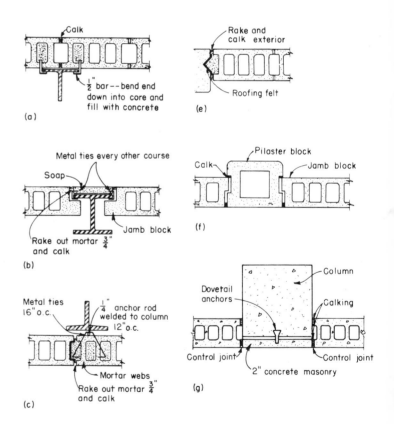

Fig. 4-31. Control joints at columns and pilasters.

tural framing members, such as columns and pilasters, a space should be allowed between the masonry and member to allow free movement. A control joint should be located at the column or pilaster (Fig. 4-31).

When a concrete masonry wall is reduced in thickness across the face of a column, a control joint should be placed along one or both sides of the column. Thin concrete masonry across the column face should be tied to the column by means of dovetail anchors (Fig. 4-31g) or another suitable device.

93

Where bond beams are provided only for crack control, control joints should extend through them. If there is a structural reason for a bond beam, a dummy groove or raked joint should be provided to control the location of the anticipated crack.

A concrete masonry or cast-in-place concrete foundation having both sides backfilled does not usually require control joints. However, long concrete masonry basement walls may require control joints, continuous metal ties (joint reinforcement), or reinforcing bars.

Where concrete masonry units are used as a backup for another material with masonry bond, the control joints should extend through the facing. Control joints need not extend through the facing when using flexible bonds (metal ties).

Control joints should extend through plaster applied directly to concrete masonry units. Plaster applied on lath that is furred out from concrete masonry requires control joints over previous joints in the base.

Joint Reinforcement

Although concrete masonry walls can be built essentially free of cracks, it is the infrequent crack for which joint reinforcement (continuous metal ties) is provided. The function of joint reinforcement is not to eliminate cracking in concrete masonry walls but merely to prevent the formation of conspicuous shrinkage cracks. Joint reinforcement does not be-

come effective until the concrete masonry begins to crack. At this time the stresses are transferred to and redistributed by the steel. The result is evenly distributed, very fine cracks hardly visible to the naked eye.

The effectiveness of joint reinforcement depends on the type of mortar and the bond between the mortar and the longitudinal wires. The better the bond strength, the more efficient the reinforcement in arresting any cracking. In-service experience has shown that only Types M, S, and N mortar should be considered for use with joint reinforcement.

After the joint reinforcement is placed on top of the bare masonry course, the mortar is applied to cover the face shells and joint reinforcement. Minimum recommended mortar cover of the wire is 5/8 in. for the exterior wall face and 1/2 in. for the interior face, as shown previously in Fig. 4-5c.

Prefabricated or job-fabricated corner and T-type joint reinforcement should be used around corners and to anchor abutting walls and partitions (Fig. 4-12d and e). Prefabricated corners and tees are considered superior because they are more accurately formed, fully welded, and easier to install. A 6-in. lapping of side wires at splices is essential to the continuity of the reinforcement so that tensile stress will be transmitted.

As can be seen in Table 4-1, the vertical spacing of joint reinforcement is interdependent with the spacing of control joints. In addition, joint reinforcement should be located as follows:

1. In the first and second bed joints immediately above and below wall openings. The reinforce-

Fig. 4-32. Wall with scored block. Joint reinforcement controls drying shrinkage and temperature movements.

ment should extend not less than 24 in. past either side of the opening or to the end of the panel, whichever is less.

2. In the first two or three bed joints above floor level, below roof level, and near the top of the wall.

Joint reinforcement need not be located closer to a bond beam than 24 in. It should not extend through control joints unless specifically called for and detailed in the plans.

Layout of Structural Features

Modular Planning

Modular planning is a method of coordinating the dimensions of various building components to simplify the work and thus lower the cost of construction. It minimizes cutting and fitting of units on the job, operations that slow up construction. In a modular plan for concrete masonry construction, all horizontal dimensions are given in multiples of half the nominal length of a concrete block, usually 8 in. Vertically the dimensions are given in multiples of the full nominal height of the block.

Tables 4-2 and 4-3 give modular lengths and heights for walls. If necessary, head and bed joints may have different thicknesses.

Table 4-2. Length of Concrete Masonry Walls by Stretchers

No. of stretchers	Wall length*
1	1'4"
1-1/2	2'0"
2	2'8"
2-1/2	3'4"
3	4'0"
3-1/2	4'8"
4	5'4"
4-1/2	6'0"
5	6'8"
5-1/2	7'4"
6	8'0"
6-1/2	8'8"
7	9'4"
7-1/2	10'0"
8	10'8"
8-1/2	11'4"
9	12'0"
9-1/2	12'8"
10	13'4"
10-1/2	14'0"
11	14'8"
11-1/2	15'4"
12	16'0"
12-1/2	16'8"
13	17'4"
13-1/2	18'0"
14	18'8"
14-1/2	19'4"
15	20'0"
20	26'8"

*Based on units 15-3/8 in. long and half units 7-5/8 in. long, with 3/8-in.-thick head joints.

Table 4-3. Height of Concrete Masonry Walls by Courses

No. of courses	Wall height					
	3/8-in. bed joint		7/16-in. bed joint		1/2-in. bed joint	
	8-in. block	4-in. block	8-in. block	4-in. block	8-in. block	4-in. block
1	8"	4"	8-1/16"	4-1/16"	8-1/8"	4-1/8"
2	1'4"	8"	1'4-1/8"	8-1/8"	1'4-1/4"	8-1/4"
3	2'0"	1'0"	2'0-3/16"	1'0-3/16"	2'0-3/8"	1'0-3/8"
4	2'8"	1'4"	2'8-1/4"	1'4-1/4"	2'8-1/2"	1'4-1/2"
5	3'4"	1'8"	3'4-5/16"	1'8-5/16"	3'4-5/8"	1'8-5/8"
6	4'0"	2'0"	4'0-3/8"	2'0-3/8"	4'0-3/4"	2'0-3/4"
7	4'8"	2'4"	4'8-7/16"	2'4-7/16"	4'8-7/8"	2'4-7/8"
8	5'4"	2'8"	5'4-1/2"	2'8-1/2"	5'5"	2'9"
9	6'0"	3'0"	6'0-9/16"	3'0-9/16"	6'1-1/8"	3'1-1/8"
10	6'8"	3'4"	6'8-5/8"	3'4-5/8"	6'9-1/4"	3'5-1/4"
15	10'0"	5'0"	10'0-15/16"	5'0-15/16"	10'1-7/8"	5'1-7/8"
20	13'4"	6'8"	13'5-1/4"	6'9-1/4"	13'6-1/2"	6'10-1/2"
25	16'8"	8'4"	16'9-9/16"	8'5-9/16"	16'11-1/8"	8'7-1/8"
30	20'0"	10'0"	20'1-7/8"	10'1-7/8"	20'3-3/4"	10'3-3/4"
35	23'4"	11'8"	23'6-3/16"	11'10-3/16"	23'8-3/8"	12'0-3/8"
40	26'8"	13'4"	26'10-1/2"	13'6-1/2"	27'1"	13'9"
45	30'0"	15'0"	30'2-13/16"	15'2-13/16"	30'5-5/8"	15'5-5/8"
50	33'4"	16'8"	33'7-1/8"	16'11-1/8"	33'10-1/4"	17'2-1/4"

sions of the finished door and window openings are 3/8 in. greater than their modular dimensions given on the plan. The actual dimension of the finished wall is 3/8 in. less than its modular dimension on the plan. However, the concrete foundation is built to the full modular dimension and theoretically the mason starts the corner masonry unit 3/16 in. in from the end.

Of course, modular design for concrete masonry requires that window and door frames be of the same mode, as shown in Figs. 4-35 and 4-36. The shaded portion of Fig. 4-35 indicates the cutting of units required by nonmodular openings and nonmodular wall length.

Fig. 4-33. Modular planning minimizes the cutting and fitting of block on the job.

Door and Window Openings

The example of modular planning given in Fig. 4-34 shows the widths of door and window openings as well as wall lengths in multiples of 8 in. Since the concrete block is produced with dimensions 3/8 in. less than its nominal or modular length of 16 in., the actual dimen-

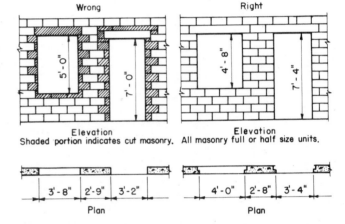

Fig. 4-35. Examples of wrong and right planning of concrete masonry wall openings based on 8×8×16-in. block.

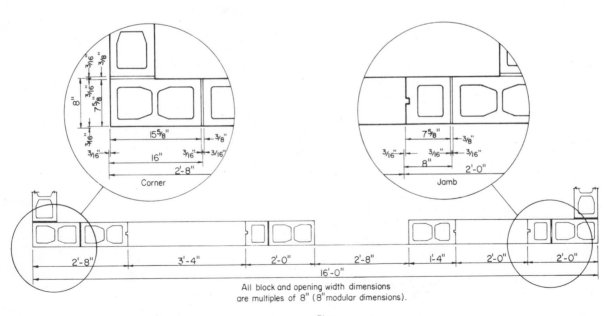

All block and opening width dimensions
are multiples of 8" (8" modular dimensions).

Plan

Fig. 4-34. Modular planning of a wall.

Corners

An important consideration in modular planning is the method to be adopted for constructing corners. Eight-inch-thick walls do not pose a problem in this regard, but thicker or thinner walls require some attention so that the 4- or 8-in. module is preserved. Figs. 4-38 through 4-42 show some suggested details for handling corner layouts for walls of various thicknesses.

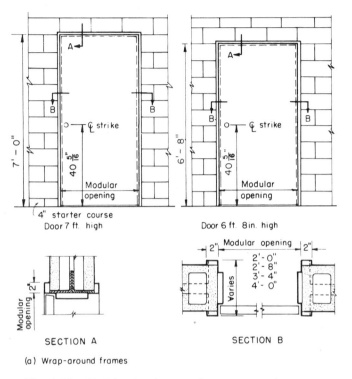

(a) Wrap-around frames

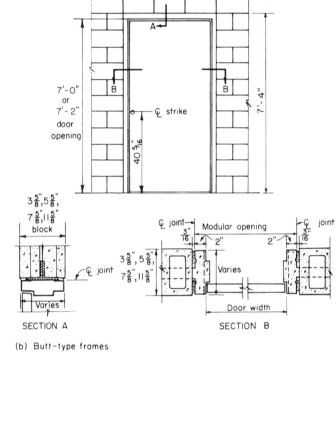

(b) Butt-type frames

Fig. 4-36. Modular-size door openings.

Fig. 4-37. Screen wall enclosing a refuse area. Note the neat corners.

97

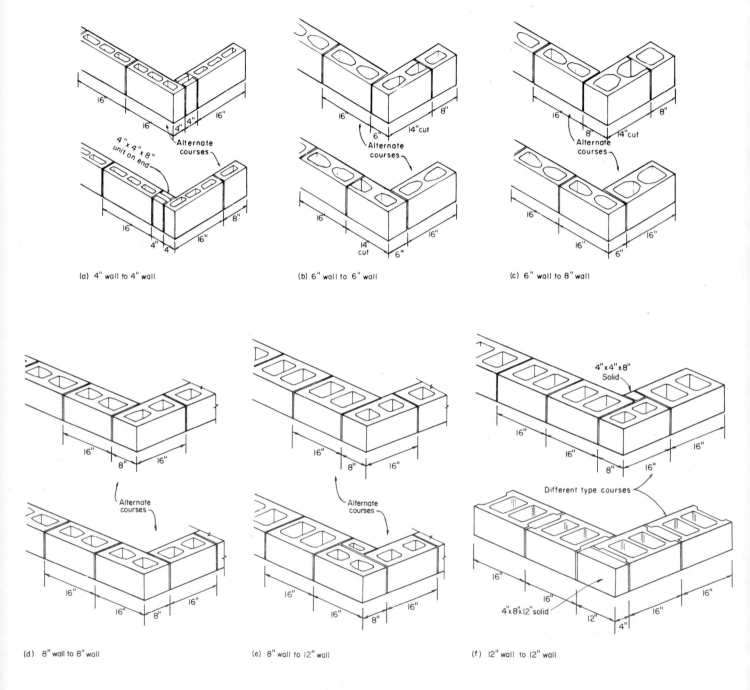

(a) 4" wall to 4" wall

(b) 6" wall to 6" wall

(c) 6" wall to 8" wall

(d) 8" wall to 8" wall

(e) 8" wall to 12" wall

(f) 12" wall to 12" wall

Fig. 4-38. Standard corner layouts for walls 4, 6, 8, and 12 in. thick.

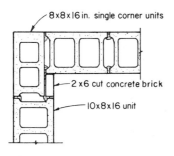

8x8x16 in. single corner units

2 x 6 cut concrete brick

10x8x16 unit

Standard corner construction

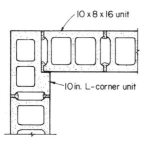

10 x 8 x 16 unit

10 in. L-corner unit

Special L-corner construction

(a) 10" wall to 10" wall

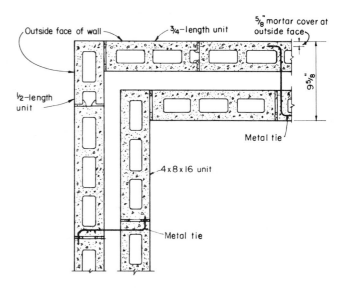

Outside face of wall

¾-length unit

⅝" mortar cover at outside face

9⅝"

Metal tie

½-length unit

4 x 8 x 16 unit

Metal tie

Fig. 4-40. Corner layout for 10-in. cavity wall.

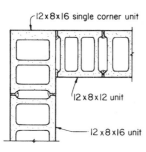

12x8x16 single corner unit

12 x 8 x 12 unit

12 x 8 x 16 unit

Standard corner construction

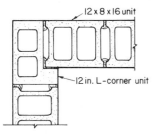

12 x 8 x 16 unit

12 in. L-corner unit

Special L-corner construction

(b) 12" wall to 12" wall

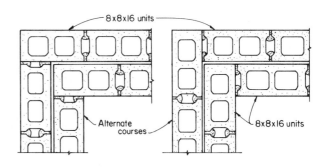

8 x 8 x 16 units

Alternate courses

8x8x16 units

(a) 8" units

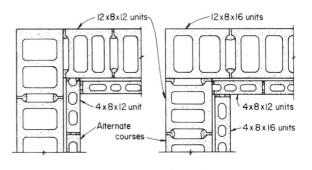

12 x 8 x 12 units

12 x 8 x 16 units

4 x 8 x 12 unit

Alternate courses

4 x 8 x 12 units

4 x 8 x 16 units

(b) 12" and 4" units

Fig. 4-39. Corner layouts comparing standard and special L units for 10- and 12-in. walls.

Fig. 4-41. Corner layouts for 16-in. walls.

99

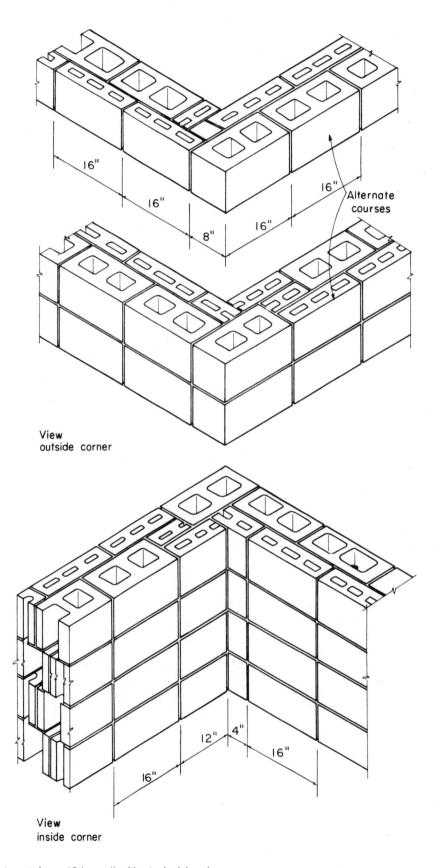

View
outside corner

View
inside corner

Fig. 4-42. Corner layout for a 12-in. wall with stacked bond pattern.

Intersections

Connection of intersecting walls needs planning unless the method is predetermined by a building code. If there is a choice, two items must be studied: whether the wall will require lateral support and where the control joints will be located. Several conditions are illustrated in Fig. 4-43.

Intersecting load-bearing block walls that depend upon one another for continuity and lateral support, as at A in Fig. 4-43, should be securely anchored to resist all forces that might tend to separate them. Such walls should be connected with true masonry bond so that half of the units of each wall are embedded in the other wall. Alternate connections are shown in Fig. 4-44. Another method is to use the detail of Fig. 4-45a, except that the elastic joint sealant is replaced with a regular mortar joint, as in Fig. 4-46.

At B or C in Fig. 4-43, a full control joint is assumed to be necessary to allow movement in two directions. The joint would be calked, contain no mortar, and have no block across it.

At the D joint in Fig. 4-43, the wall running north-south will try to move transversely to the other. In addition, it requires lateral support—a steel tiebar used as shown in Fig. 4-45a. The appropriate cores are filled with mortar after pieces of metal lath are placed under the cores to support the filling. The ends of the tiebar are embedded in this mortar filling. The result is a hinged control joint that permits the wall to move slightly (at right angles to the abutting wall) and yet

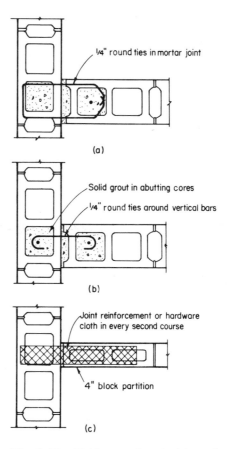

Fig. 4-44. Rigid connections for intersecting walls.

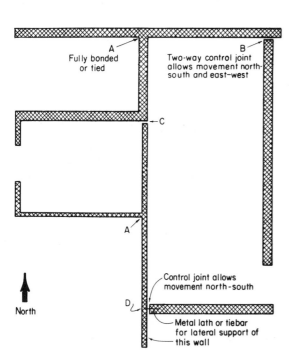

Fig. 4-43. Jointing at intersecting walls.

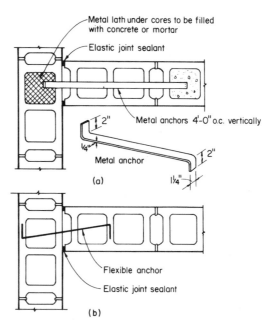

Fig. 4-45. Flexible connections for intersecting walls.

Fig. 4-46. Steel tiebar provides lateral support to wall at right.

have lateral support. The tiebars are spaced not over 4 ft. apart vertically.

In another method for D a control joint is placed at the junction of the walls. Lateral support is provided by joint reinforcement, strips of metal lath, or 1/4-in.-mesh galvanized hardware cloth placed across the joint between the two walls (Fig. 4-44c). When metal strips are used, they are placed in alternate courses in

Fig. 4-47. Hardware cloth provides lateral support to wall at rear.

the wall or not more than 18 in. on center. They are sufficiently flexible to permit lateral movement of the abutted wall. If one wall is constructed first, the metal strips are built into it and later tied into the mortar joint of the second wall.

Bond Beams

Bond beams are reinforced courses of block that bond and integrate a concrete masonry wall into a stronger unit. They increase the bending strength of the wall and are particularly needed to resist high winds and hurricane and earthquake forces. In addition, they exert restraint against wall movement and thus reduce the formation of cracks. They may even be used at vertical intervals instead of vertical reinforcing steel.

The bond beams are constructed with special-shape masonry units (Fig. 1-11c through f, page 14) filled with concrete or grout and reinforced with embedded steel bars. Bond beams are usually located at the top of walls to stiffen them. Since they have appreciable structural strength, they can be located to serve as lintels over doors and windows. When bond beams are located just above the floor, they act to distribute the wall weight (making the wall a deep beam) and thus help avoid wall cracks if the floor sags. Bond beams

Fig. 4-48. Construction of a bond beam for wall reinforcement.

may also be located below a window sill. Examples of common bond-beam locations are shown in the Appendix.

In load-bearing walls, bond beams are best placed in the top course immediately beneath the roof framing system. Where heads of openings occur within 2 ft. of the roof framing system and/or ceiling in multistory buildings, a bond beam at or immediately above the lintel can be considered the equivalent of a bond beam at the top of the wall or at ceiling height.

In non-load-bearing walls, bond beams may be placed in any one of the top three courses below the roof slab or deck. This permits the bond beam to serve three functions—as a structural tie, lintel, and crack control device.

Reinforcement of bond beams must satisfy structural requirements but should not be less than two No. 4 steel bars in 8-in.-wide bond beams and two No. 5 steel bars in 10- and 12-in.-wide bond beams. In some earthquake-prone areas building codes require a 16-in.-deep bond beam with two additional No. 4 bars located in the top of the beam. Bars should be bent around corners and lapped (Fig. 4-49) according to the local building code.

When the bond beams serve only as a means of crack control, they should be discontinuous at control joints. Where structural considerations require that bond beams be continuous across control joints, a dummy groove should be provided to control the location of the anticipated crack.

If the bond beams are used to replace joint reinforcement, their spacings should be as given in Table 4-4. Note that the area of steel required in the bond beams is greater than that required in the joint reinforcement. This is due not only to a lower reinforcing bar yield strength, but also to a loss of bond-beam effectiveness (assumed to be reduced one-third) because of the wetting effect of grout on the wall and the accompanying increase in ultimate drying shrinkage of the wall.

Lintels

Lintels are reinforced horizontal members that span openings in walls; they function as beams to support the weight of the wall and other loads over openings. Concrete masonry lintels may be made up of specially formed lintel masonry units (Fig. 1-11g, page 14), bond-beam masonry units, or standard units with depressed, cut-out, or grooved webs. The units are laid end to end to form a channel for placement of reinforcing steel and grout. These lintels are easy to construct, do not require heavy hoisting equipment, and match the bond pattern and surface texture of the surrounding masonry. Lintels may also be conventional precast concrete, steel angles, or concrete masonry members molded on special machines that produce surface textures similar to block.

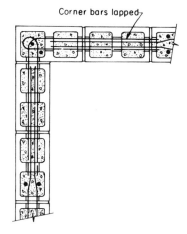
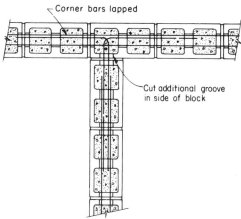

Fig. 4-49. Bond-beam corner details.

Table 4-4. Equivalent Spacing of Bond Beams and Joint Reinforcement*

Maximum spacing of joint reinforcement,** in.	Spacing based on bond-beam reinforcement†		
	Two No. 4 bars	Two No. 5 bars	Two No. 6 bars
24	8'0"	8'0"	8'0"
16	5'4"	8'0"	8'0"
8	2'8"	4'8"	6'8"

*Adapted from Table 3-3, Ref. 59.
**No. 9 gage wire with yield strength of 65,000 psi.
†Yield strength of 40,000 psi.

Lintels should have a minimum bearing of 6 in. at each end. A rough rule of thumb is to provide 1 in. of bearing for every foot of clear span.

Sills

Some concrete masonry producers make sill units in a modular length of 7-5/8 in., as shown in Fig. 4-50. They are mortared together to form a continuous sill. Because of the difficulty in ensuring a watertight mortar joint on top of the sill, the course below is constructed of solid units or a bond beam.

Other sill units are shown in Fig. 1-11a (page 14).

Piers and Pilasters

Piers are isolated columns of masonry while pilasters are columns or thickened wall sections built contiguous with and forming part of a masonry wall. Pilasters may project on one or both sides of the wall. If the projection is entirely on one side, the pilaster is generally referred to as a flush pilaster or as an interior or exterior pilaster.

Both piers and pilasters are used to support heavy, concentrated vertical roof or floor loads and provide lateral support to the walls. They also offer an economic advantage by permitting construction of higher and thinner walls with plain concrete masonry; otherwise, the walls would have to be reinforced or made thicker.

Piers and pilasters may be constructed of special concrete masonry units (Figs. 1-14 and 1-15, pages 16-17) or units similar to those used in the wall,

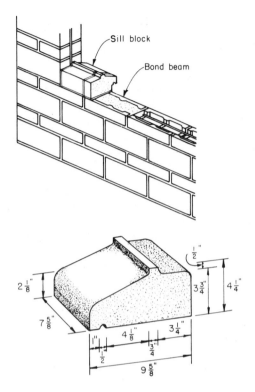

Fig. 4-50. Sill block of modular dimensions.

Fig. 4-51. Piers and pilasters accent a wall built of split block laid in an offset bond pattern.

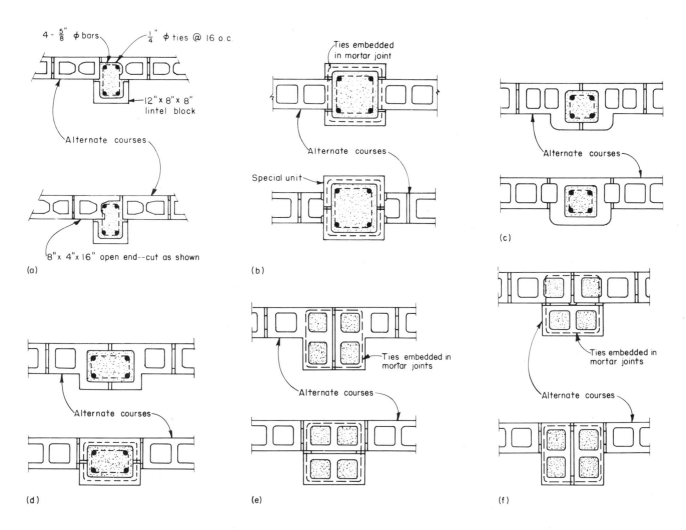

Fig. 4-52. Pilaster layouts.

whether solid or hollow. Hollow units may be grouted and may or may not contain embedded reinforcement. Grouted piers and pilasters have all vertical joints fully mortared. Pilaster units should be laid with the same mortar quality as is planned for the masonry between pilasters.

Some typical pilaster designs are shown in Figs. 4-52 and 4-53. Most of these layouts are adaptable to piers. Note that grouted piers and pilasters require ties embedded in the face-shell mortar bedding. These ties are necessary to hold the masonry units together while the fresh grout exerts fluid pressure, and to make the units work together. Tie diameter should not exceed half the joint thickness.

If grouted units are reinforced, it is recommended that they also contain 1/4-in. ties with closed (lapped) ends. These ties should be in contact with the outside of the vertical reinforcing bars. For pilasters such as those in Figs. 4-52 and 4-53, the 1/4-in. ties are installed as the block work progresses.

Planning Weathertight Walls

The outside wythe of a cavity wall can be considered as a rain screen—as are veneered and shingled walls that are backed up with a vented air space. Such walls are highly weather-resistant even in driving rains. On the other hand, solid masonry construction is more vulnerable to leakage. In choosing this type of construction, the designer and builder must accept the possibility and consequences of some leakage or else they must use greater care in selecting materials and overseeing installation. While workmanship is the most important element, it isn't fair to always hold the mason responsible for leaks due to poor workmanship. The owner, architect, and builder share the responsibility because they govern the type of workmanship desired.

The elements of watertight concrete masonry walls

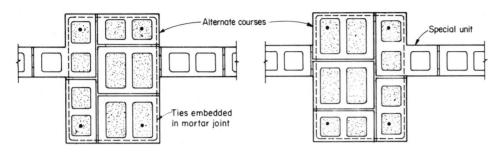

(a) Offset pilaster 24"x 32"

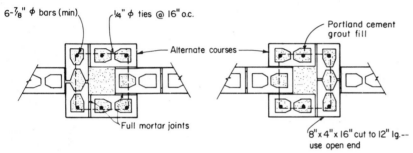

(b) Centered pilaster 24"x 24"

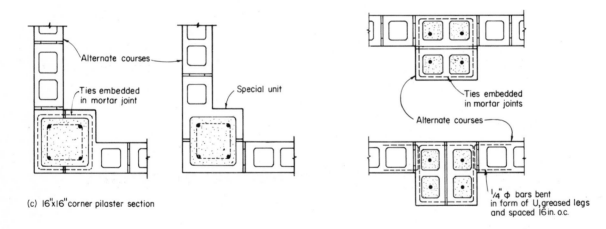

(c) 16"x16" corner pilaster section

(d) Unbonded pilaster 16"x16"

Fig. 4-53. Other pilaster layouts.

are discussed throughout this handbook. They include sound masonry units, proper mortar, and a high standard of construction. Moisture transmission is viewed by the CSA* as follows:

(i) Rain water can penetrate masonry in several ways: through separation (lack of bond between the mortar and the masonry units), or through shrinkage cracks in the mortar or masonry units. It is also recognized that, under certain conditions, intense wind-blown rain, over a prolonged period, will penetrate some masonry units and mortars;

(ii) The ability to exclude rain depends upon the type and design of the wall, in addition to the wall thickness, materials,

and workmanship. Solid masonry can be relied upon to give completely satisfactory protection against rain penetration, when correct design and proper workmanship considerations have been undertaken to counteract this factor;

(iii) A substantial addition to the resistance of solid masonry to rain penetration is afforded by back-parging and completely filling all mortar joints, including the vertical collar joint between the masonry wythes. In contrast with solid masonry, a soundly designed and constructed cavity wall provides superior resistance to rain penetration.

*Ref. 54, page 22.

ASTM and CSA specifications for concrete brick and block do not specify water permeability, but they do note that protective coatings may be required to prevent water penetration.

Portland cement plaster (stucco) will generally suffice to make a hollow concrete masonry wall weathertight. Acrylic paints and other coatings may also do this, but they may have to be reapplied every few years. Additional information on plaster, paints, and other coatings is given in Chapter 7.

Leaky walls are not confined to any one type of masonry construction. Leaks can occur in walls built of the best materials. The percentage of those that leak is small but receives a disproportionate amount of attention.

Flashing

It is difficult to completely prevent rainwater from entering walls at parapets, sills, projections, recesses, roof intersections, etc., unless proper flashing is installed. In areas subjected to severe driving rains or where experience has shown that water penetration is to be expected, special flashing and weepholes should be provided. Design details of flashing and weepholes are shown in the Appendix.

Any moisture that enters a cavity wall will gradually travel downward. To divert this water to the exterior of the building, continuous flashing and weepholes are installed at the bottom of a cavity (Fig. A-12, page 194). Flashing is not necessary where there is no basement, and where the bottom of the cavity is above grade and several inches below the lowest bearing level of the first floor.

Where there is a basement and the floor consists of wood joists, the flashing may be located above the bottom of the joists (Fig. A-25, page 198). If metal flashing is used, it may be extended at least 2 in. past the inside face of the wall and bent downward at an angle to serve as a termite shield. If the flashing is not required to serve as a termite shield, it may be stopped 1/2 in. from the outside faces of the wythes. In concrete slab-on-ground construction, the flashing extending into the interior wythe may be above the top level of the slab (Fig. A-2, page 192).

Flashing should be installed over all windows, doors, and other wall openings not completely protected by overhanging projections. Although flashing may not be required under monolithic sills, it is advisable if sill block are used, unless the course under the sill is solid block or a bond beam (Fig. 4-50). Both ends of the sill flashing should be extended beyond the jamb line and turned up at least 1 in. into the wall. Where the underside of the sill does not slope away from the wall or where no drip is provided, the flashing should be extended and bent down to form a drip. Otherwise, water running down windows and over sills will continue down the face of the building and probably cause unsightly stains.

In structural frame buildings the inner wythe of a cavity wall is constructed flush with and anchored to the beams and columns, while the outer wythe is supported by a steel shelf angle attached to a spandrel beam at each floor level (Fig. A-14b, page 195). Flashing normally is not necessary when galvanized or stainless steel angles are used, except to cover the joints between the lengths of angles; in that case flashing should be placed on the shelf angle and extended at least 8 in. up and over the beam or anchored into a reglet in the beam.

One detail used for flashing at parapets is shown in Fig. 4-54. Others appear in Figs. A-18, A-20, and A-22, pages 196-197.

Suitable flashing materials must be: (1) impervious to moisture penetration; (2) resistant to corrosion caused by exposure either to the atmosphere or to the caustic alkalies that may be present in mortar; (3) sufficiently tough to resist puncture, abrasion, or other damage during installation; and (4) easily formed to the desired shape and capable of retaining this shape throughout the life of the structure. The choice of material is governed mainly by cost and suitability. It is advisable to select the type of flashing material carefully since repair and replacement costs will be much higher than the original cost.

Materials generally used for flashing are copper, stainless steel, bituminous fabrics, and plastics. Copper, a durable and easily workable material, has an excellent performance record but is more costly than most other flashing materials. It is also available in special preformed shapes. It does not react with fresh mortar unless chlorides are present. When copper is exposed to weather, rainwater runoff may stain or discolor the masonry surfaces below. Where this staining or discoloration is objectionable, coated copper should be used for flashing.

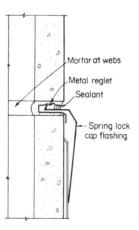

Fig. 4-54. Metal reglet used for flashing at parapets.

Stainless steel is durable, highly resistant to corrosion, and workable. Stainless steel flashing is available in several gages and finishes. It will not stain adjacent areas and resists rough handling.

Bituminous fabrics are less costly but also less durable than copper or stainless steel flashing. Care must be exercised during their installation in order to avoid tears and punctures.

Flashing made of plastic materials is also available. However, not all plastics are suitable for use in contact with mortar and thus it is necessary to rely on the past performance of a particular material before selecting it for use in a concrete masonry wall.

Combination flashing consists of materials combined to utilize the best properties of each effectively. Examples of combination flashing have plastic- or asphalt-coated metals, steel- or fiberglass-reinforced bituminous fabrics, and copper-plated stainless steel.

Flashing will reduce the flexural strength of a wall by reducing its continuity in bending resistance and shear. This is not an important factor for houses and small buildings. However, for buildings with tall or thin walls it must be taken into account by the structural designer.

Weepholes

Weepholes should be provided immediately above all flashing or other waterstops to drain away any accumulated water (Fig. A-2, page 192). The holes are usually located in the head joints of the outer wythe and spaced about 2 ft. apart. In no case should weepholes be located below grade. They should also be kept small to exclude rodents.

Weepholes are formed by: (1) omitting mortar from part or all of a joint, or (2) placing short lengths of greased or oiled inserts (such as rods, tubing, hose, or cord) into the mortar and extracting them when the mortar is ready for tooling. The inserts should extend up into the cavity for several inches to provide a drainage channel through any mortar droppings that might have accumulated.

Whenever possible, the cavity side of weepholes should be covered with copper or plastic insect screen cloth to prevent the entry of insects. Material such as fibrous glass may also be placed into the open weepholes. Sometimes absorbent inorganic material is inserted into the holes to act as "wicks," drawing moisture out of the cavity. This is especially recommended over lintel or spandrel flashing to prevent the likelihood of stain on the wall below. Weepholes filled with inorganic materials should be spaced not more than 16 in. on centers.

Safeguards Against Hurricanes and Earthquakes

To safeguard against high winds, hurricanes, and earthquakes (Seismic Zones 2 and 3), HUD requires that one- and two-family masonry dwellings be tied together with reinforcing steel.* Concrete masonry walls should contain reinforcing bars that run horizontally and vertically, extending from the footing to a bond beam at the eave level. Not only should the bars be anchored in the footing and the bond beam, but the cores should be filled in with grout. In addition, roof trusses should be tied down into the bond beam with hurricane clips or steel anchors (Fig. A-23, page 198).

---

*See Ref. 67, Sec. 604-1.2.

Chapter 5

ALL-WEATHER CONCRETE MASONRY CONSTRUCTION

The key to successful and satisfactory construction of concrete masonry in any weather—hot or cold—lies in advanced planning and satisfactory preparation. All-weather construction involves some change in procedures and additional equipment and supplies. The need for these must be anticipated if construction is to be continuous and profitable.

Both hot and cold weather significantly influence the entire masonry construction industry. Hot-weather problems often have been encountered but not recognized, resulting in some sacrifice of quality or increase in construction costs. On the other hand, greater extension of the construction season into the winter months in recent years has resulted in better utilization of manpower and brought to the forefront some techniques of construction not previously well known.

An important part of planning for all-weather construction is accurate weather-forecasting. Builders can plan their construction on the basis of their own weather experience plus information available from the weather bureau. Weather factors important to concrete masonry construction include temperature, wind, rain, snow, humidity, and cloudiness. Combinations of these factors affect construction workers and materials much more seriously than any single factor.

For example, wind and temperature together create a greater impact or chill factor than temperature alone. The cooling effect of a 20-mph wind at 20 above zero (deg. F.) is the same as that of still air at 10 below. Furthermore, a combination of high temperature, low relative humidity, and high wind can cause the early drying of mortar much more rapidly than can one of these elements alone.

Although "normal," "cold," and "hot" are relative terms for masonry construction, "normal" is generally considered as any temperature between 40 and 90 deg. F. Building codes and specifications vary somewhat in this respect. In any case, it should be remembered that some problems may be experienced with such temperatures; for example, those associated with hot weather may occur even when the temperature is below 90 deg. F.

With modifications of design and construction procedures, concrete masonry construction can be completely satisfactory despite the weather. In many cases concrete masonry construction during hot weather may be little, if any, more expensive than the same construction at normal temperatures; the added cost of masonry construction due to cold weather often amounts to less than 1.5%. As the departure from normal becomes greater, however, the measures necessary to overcome the effects of temperature become more important and more costly.

Hot-Weather Construction

Hot weather poses some special problems for concrete masonry construction. These arise, in general, from higher temperatures of materials and equipment and more rapid evaporation of the water required for cement hydration and curing. Other factors contributing to the problems include wind velocity, relative humidity, and sunshine.

Masonry Performance at High Temperatures

As the temperature of mortar increases, there are several accompanying changes in its physical properties:

1. Workability is lessened; that is, for a given workability, more water is required.
2. A given amount of air-entraining agent will yield less entrained air.
3. Initial and final set will occur earlier while evaporation will generally be faster.
4. Depending on the surface characteristics, temperature, and moisture content of the concrete masonry units, their suction of moisture from the mortar will be faster.

Fig. 5-1. Slump block facing used in an arid region where hot-weather construction procedures are regularly practiced.

The result of these changes is that mortar will rapidly lose water needed for hydration. Despite its higher initial water content, mortar will be somewhat more difficult to place and the time available for its use will be shorter.

Early surface drying of mortar joints is particularly harmful. Evaporation removes moisture more rapidly from the outer surface of mortar joints, but the inner parts retain moisture longer and so develop greater strength. A difference of strength across the thickness of the wall reduces the buckling strength of a wall that is concentrically loaded. Also, weak mortar on the surface reduces the strength of the wall under wind and other horizontal loads.

Selection and Storage of Materials

During hot weather there is a temptation to reduce the amount of cementitious material in the mortar mixture in order to lessen the heat of hydration released at early ages. Actually, the better solution is to *increase* the amount of cementitious material. This will accelerate rather than retard the mortar's gain in

early strength and thus secure maximum possible hydration before water is lost by evaporation.

Mortar materials stored in the sun can become hot enough to significantly affect the temperature of the mortar mixture itself. Covering or shading such materials from the sun can be helpful. For example, sand delivered to jobsites normally contains free moisture ranging from 4 to 8%, which is sufficient to ensure that a covered or shaded stockpile of sand remains reasonably cool. Of course, if the moisture content drops much below this level, the stockpile should be sprinkled to increase evaporative cooling. When evaporating, 1 gal. of water will cool 1 cu.yd. of sand 20 deg. F. at the surface of a stockpile.

In hot weather the main objective is to see that *all* of the materials of concrete masonry are placed without having acquired excess heat. That is, heat should be minimized in concrete masonry units—by storing them in a cool place—and the mortar mixture should be relatively cool. The most effective way of cooling mortar during mixing is to use cool water. Immediately after mortar has been mixed, it begins to rise in temperature and must be protected from further heat gain during construction.

Other Construction Practices

Attention should be given to cooling metal equipment with which the masonry materials, particularly mortar, come into contact. Relatively cool mortar can heat rapidly when transported in a metal wheelbarrow or other container that has been exposed for hours to the sun's rays. Metal mortarboards can become quite hot and wooden ones can become very absorptive in hot weather. Flushing them with water immediately before use and/or working under sunshades can lessen such difficulties.

Since wind and low relative humidities cause increased evaporation, the use of wind screens and fog (water) sprays can effectively reduce the severe effects of hot, dry, windy weather. Also, covering walls immediately after construction will effectively slow the rate of loss of water from masonry. Damp-curing is very effective, particularly in development of tensile bond. If the wall will be subjected to flexure, consideration should be given to damp-curing.

In areas where high ambient air temperatures are common, masonry construction is sometimes rescheduled to avoid hot, midday periods. Construction at night or during the early morning hours can avoid many hot-weather problems.

Cold-Weather Construction

When the ambient temperature falls below normal (40 deg. F.), the productivity and workmanship of masons and the performance of materials may be lowered.

During cold weather masons are concerned not only with their normal construction tasks but also with personal comfort, additional materials preparation and handling, and protection of structures. As temperatures continue to drop, these extra activities consume more time.

Masonry Performance at Low Temperatures

Immediately after concrete masonry units are laid during cold weather, several factors come into play. The absorptive masonry units tend to withdraw water from mortar, but mortar, having the property of retentivity, tends to retain water. The surrounding air may chill masonry as well as withdraw water through evaporation. Also, if the masonry units are cold when laid, they will drain heat from mortar. Any combination of these factors influences strength development.

As the ambient temperature falls below normal, mortar ingredients become colder and the heat-liberating reaction between portland cement and water is substantially reduced. Hydration and strength de-velopment are minimal at temperatures below freezing. However, construction may proceed at temperatures below freezing if the mortar ingredients are heated. As the ambient temperature decreases, the masonry units should be heated and the structure maintained above freezing during the early hours after construction.

Mortars mixed with cold but unfrozen materials possess plastic properties quite different than those at normal temperatures. The water requirements for a given consistency decrease as the temperature falls; more air is entrained with a given amount of air-entraining agent; and initial and final set are delayed. Also, with lower temperature, the strength gain of mortar is less, although final strength may be as high or higher than that of mortar used and cured at more normal temperatures.

Heated mortar materials produce mortars with performance characteristics identical to those in the normal-temperature range, and thus heating is desirable for cold-weather masonry construction. Mortars mixed to a particular temperature and subjected to a lower ambient level lose heat until they reach the ambient temperature. If the ambient temperature is below freezing when the mortar temperature reaches 32 deg. F., the mortar temperature remains constant until all water in the mortar is frozen.* Afterward, the mortar temperature continues to descend until it reaches the level of the ambient temperature.

The rate at which masonry freezes is influenced by the severity of air temperature and wind, the temperature and properties of masonry units, and the temperature of mortar. When fresh mortar freezes, its performance characteristics are affected by many factors: water content, age at freezing, strength development prior to freezing, etc. Frozen mortar takes on all the outward appearances of hardened mortar, as evidenced by its ability to support loads as well as its ability to bond to surfaces.

Mortar possessing a high water content expands when it freezes—and the higher the water content, the greater the expansion. The expansive forces will not be disruptive if moisture in the freezing mortar is below 6%. Therefore, every effort should be made to achieve mortar with low water content. Dry masonry units and protective coverings should be used.

Mortar that is allowed to freeze gains very little strength and some permanent damage is certain to occur. If the mortar has been frozen just once at an early age, it may be restored to nearly normal strength by providing favorable curing conditions. However, such mortar is neither as resistant to weathering nor as watertight as mortar that has never been frozen.

---

*The loss of 144 Btu is required to change 1 lb. of water at 32 deg. F. to 1 lb. of ice at 32 deg. F.; on the other hand, only about 1/2 Btu is lost in lowering the temperature of 1 lb. of *ice* 1 deg. F. (For 1 lb. of *water*, 1 Btu is lost in lowering the temperature 1 deg. F.)

Selection of Materials

Cold-weather concrete masonry construction generally requires only a few changes in the mortar mixture. Concrete masonry units used during normal temperatures may be successfully used during cold weather. Under the prevailing recommendations for winter construction (Table 5-1), the masonry units will generally lower the moisture within the mortar to below 6% and so any subsequent accidental freezing will not be disruptive.

At low temperatures mortar performance can be improved—with an early strength gain—by use of Type III or High-Early-Strength cement. Also, mortar made with lime in the dry, hydrated form is preferred to slaked quicklime or lime putty because it requires less water.

Admixtures often considered for inclusion in mortar are antifreezes, accelerators, corrosion-inhibitors, air-entraining agents, and color pigments. Those used with proven success in cold weather are the accelerators and air-entraining agents.

Certain admixtures for mortar are misunderstood in that they accelerate strength gain rather than lower the freezing point. So-called "antifreeze" admixtures, including several types of alcohol, must be used in great quantities to significantly lower the freezing point of mortar, but the compressive and bond strengths of masonry are also lowered. Therefore, antifreeze compounds are not recommended for cold-weather masonry construction.

The primary interest in accelerators is to increase rates of early-age strength development; i.e., to hasten hydration of portland cement in mortar. Accelerators

Table 5-1. Recommendations for Cold-Weather Masonry Construction*

Air temperature, deg. F.	Construction requirements	
	Heating of materials	Protection
Above 40	Normal masonry procedures.	Cover walls with plastic or canvas at end of workday to prevent water entering masonry.
Below 40	Heat mixing water. Maintain mortar temperatures between 40 and 120 deg. F. until placed.	Cover walls and materials to prevent wetting and freezing. Covers should be plastic or canvas.
Below 32	In addition to the above, heat the sand. Frozen sand and frozen wet masonry units must be thawed.	With wind velocities over 15 mph, provide windbreaks during the workday and cover walls and materials at the end of the workday to prevent wetting and freezing. Maintain masonry above 32 deg. F. by using auxiliary heat or insulated blankets for 16 hours after laying masonry units.
Below 20	In addition to the above, dry masonry units must be heated to 20 deg. F.	Provide enclosures and supply sufficient heat to maintain masonry enclosure above 32 deg. F. for 24 hours after laying masonry units.

*Adapted from guide specifications of the International Masonry Industry All-Weather Council (Ref. 62). This council is composed of the following organizations: the Bricklayers, Masons and Plasterers International Union of America, the Laborers' International Union of North America, Mason Contractors Association of America, National Concrete Masonry Association, Portland Cement Association, and the Brick Institute of America.

include calcium chloride, soluble carbonates, silicates and fluosilicates, calcium aluminate, and organic compounds such as triethanolamine. Aluminous cements and finely ground hydrated cements have also been advocated for acceleration.

The most commonly used accelerator in concrete is calcium chloride. However, its use in mortar is controversial because of possible adverse side effects, such as increased shrinkage, efflorescence, and corrosion of embedded metal. Since calcium chloride may produce corrosion failure, it should not be permitted in mortar for concrete masonry containing metal ties, anchors, door bucks, or joint reinforcement. If these factors are not involved, it is recommended that the amount of calcium chloride used should not exceed 2% by weight of portland cement or 1% by weight of masonry cement, added in solution form.

Some proprietary admixture compounds have been modified to contain corrosion-inhibitors for winter construction of concrete containing embedded metal. Although reports have been published on the performance of soluble chromates and sodium benzoate as corrosion-inhibiting compounds, their value for cold-weather concrete masonry construction has not been fully determined. Therefore, corrosion-inhibitors are not recommended.

Air-entraining agents may be added at the mixer to increase mortar workability and freeze-thaw durability at later ages, but their effects on mortars subjected to early freezing have not been established. In practice, however, masonry cement that entrains air and air-entraining cement perform satisfactorily in mortars used during winter construction.

Some color pigments contain dispersing agents to speed the distribution of color throughout the mortar mixture. The dispersing agents may have a retarding effect on the hydration of portland cement, and this retardation is particularly undesirable in cold-weather masonry construction. In addition, the masonry may have a greater tendency to effloresce.

Storage and Heating of Materials

At delivery time all masonry materials should be adequately protected for any exposure conditions at the construction site. During cold weather the safe storage of mortar materials can be accomplished by providing an improvised shelter (Fig. 5-2). As discussed later in this chapter, shelters may be erected by using scaffolding sections, enclosure covers, and lumber. With a properly erected shelter, mortar materials may be delivered, stored, protected from the elements, heated, and mixed within that shelter.

If shelters are not built, all masonry materials should be covered when the ambient temperature is below 40 deg. F. Bagged materials and masonry units should be securely wrapped with canvas or polyethylene tarpaulins and stored above the reach of moisture migrating from the ground. The masonry sand should be covered to keep out snow and ice buildup before the sand pile is heated.

Fig. 5-2. A makeshift shelter will protect not only the equipment and materials but also the mixing operation.

Fig. 5-3. Concrete block covered with tarpaulins to keep them dry.

The most important consideration in heating materials is that sufficient heat be provided to assure cement hydration in mortar. After all materials are combined, the mortar temperature should be within the range of 40 to 120 deg. F. (70 to 120 deg. F. in Canada). If the air temperature is falling, a minimum mortar temperature of 70 deg. F. becomes worthwhile. Mortar temperature in excess of 120 deg. F. may cause excessively fast hardening with a resultant loss of compressive and bond strength. Heating requirements for various air temperatures are given in Table 5-1.

Water

When the air temperature drops, water is generally the first material heated for two reasons: it is the easiest material to heat and it can store the most heat, pound for pound, of any of the materials in mortar. Recommendations vary as to the highest temperature to which water should be heated. Some specifiers put a maximum of 180 deg. F. on heating water because there is a danger of flash set if significantly hotter water comes into contact with cement. Combining sand and water in the mixer first, before adding the cement, will lower the temperature and avoid this difficulty. With this precaution and the use of aggregates that are cold enough, even boiling water may be used successfully.

Sand

When the air temperature is below 32 deg. F., sand should be heated so that all frozen lumps are thawed. Generally the temperature of the sand is raised to 45 to 50 deg. F. However, if the need exists and facilities are available, there is no objection to raising the sand temperature much higher; 150 deg. F. is a reasonable upper limit.

Sand should not be heated to a temperature that would cause decomposition or scorching. For example, when sand containing limestone or dolomite is heated to a temperature above 1,200 deg. F., carbon dioxide is liberated; free lime and magnesium oxide may then form and cause mortar contamination. When siliceous sand is heated above 1,000 deg. F., scorching can occur. A practical method is to limit the sand temperature by feel (touch of the hand). The stockpile must be mixed periodically to assure uniform heating.

A commonly used method of heating sand is to pile it over a metal pipe containing a fire (Fig. 5-5). Another method is to use steam circulated through coils or injected directly into the sand. Steam boilers are an economical source of heat for winter construction.

An ordinary 40-gal. hot-water heater will raise the temperature of 40 gal. of water about 100 deg. F. in one hour, or the temperature of 1 ton of moist, unfrozen sand about 65 deg. F. in one hour. Frozen sand or water will require more time.

Fig. 5-4. A heater pipe extends under the sand from a heater box containing a hot-water tank. Warm sand and a continuous supply of hot water are available with this method.

Fig. 5-5. Sand can be heated over fire in a pipe. Although metal culvert pipe could be used, this heater was made from 30-gal. oil drums (bottoms and tops removed) joined by tack-welding. Sand should be turned over at intervals to avoid scorching.

Masonry Units

The units can be heated by stacking them around coke- or oil-burning salamanders. Another method is to use oil, gas, or electric hot-air heaters with built-in blowers. If heated enclosures are used for construction, as discussed later, a supply of masonry units can be thawed on the scaffold (Fig. 5-6).

During very cold weather, frozen walls must be heated before grout is poured into cores or cavities. It is recommended that when the air temperature is below 40 deg. F., or it has been below 32 deg. F. during

Fig. 5-6. Concrete block are placed on the scaffold for enclosure and heating. Corrugated fiberglass panels on independent framework enclose the wall and mason's scaffold.

Fig. 5-8. One floor at a time, this high-rise building is protected from cold and inclement weather to avoid the traditional winter slowdown or shutdown in building construction. Heat and light are provided within the protected portion.

Fig. 5-7. Winter enclosure of masonry construction has many benefits for the contractor, including an earlier completion date and retention of good workers. Masons receive paychecks straight through the cold-weather season instead of only 70% of the year (northern climates).

the previous two hours, the air temperature in the bottom of the grout space should be raised above 32 deg. F. before grouting. Heated enclosures may be used for this purpose.

Construction of Temporary Enclosures

The advantages of uninterrupted construction activity during cold weather have persuaded many contractors to build temporary enclosures. Also, U.S. federal regulations now require heated enclosures on government projects to maintain year-around construction.

Providing temporary enclosures for concrete masonry construction has several objectives:

1. To achieve temperatures high enough to facilitate cement hydration in mortar and thus obtain the initial strength required for support of superimposed masonry.
2. To improve the comfort and efficiency of masons and other craftsmen.
3. To protect materials.

Since enclosures are normally built for temporary use, low-cost, lightweight, easily erected and dismantled but tightly sealed enclosure construction is the goal. The type of enclosure constructed depends on whether one or more crafts are to be protected, whether the enclosure is provided by the general contractor or the masonry subcontractor, and whether the building has single or multiple stories. Protection may be given to the walls only, the entire building, or just one floor at a time (Fig. 5-8).

Enclosure construction may use temporary, independent framework covered with sheet membranes, tarpaulins, or prefabricated panels (Fig. 5-6). Materials for enclosures include lumber, steel, canvas, building paper, and several plastics such as fiberglass and polyethylene.

The most popular sheet membrane material is polyethylene film in thicknesses of 6 to 12 mils (0.006 to 0.012 in.); it is usually clear plastic to let in the light and heat of the sun. Polyethylene reinforced with a fiber or wire mesh is widely used because it is more resistant to ripping. Panels or tarpaulins having rein-

Fig. 5-9. After the weather suddenly turned cold, this scaffolding was enclosed with polyethylene.

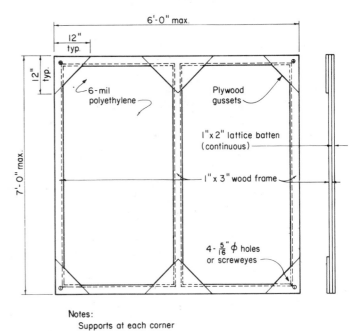

Notes:
Supports at each corner
Wind load: 15 psf
Wood bending stress: 1,200 psi + 33%

Fig. 5-10. Polyethylene enclosure panel.

forced polyethylene with grommets can be used again and again.

A common practice in concrete masonry construction is to enclose only the wall by attaching a protective cover to the mason's scaffold outside. In Fig. 5-9, when the inner wythes of the wall were laid, the enclosure was made by suspending polyethylene tarpaulins from the building framework (permanent or temporary) several feet behind the mason. Of course, prefabricated panels could have been used on the framework. They are made with 1×3- or 2×4-in. wood frames, diagonal bracing or plywood corner gussets, and stapled or nailed lattice battens to secure the membrane edges, as shown in Fig. 5-10.

The complete enclosure of a building is usually selected by the general contractor rather than the masonry subcontractor, or is required as described in the job specifications. For low-rise buildings, it offers the opportunity of protection for later craftsmen and for subsequent erection of interior concrete masonry partitions. Generally low-rise enclosure is accomplished by erecting a temporary framework of timber with polyethylene or canvas cover (Fig. 5-11).

Complete enclosure of a high-rise building begins with the first floor at ground level. Framework, walls, and first floor can be built within this enclosure. The enclosure is then lifted for construction of the second floor and the process repeated until the roof is constructed. As the enclosure is lifted, openings in the newly erected walls are closed with temporary panels until the windows and doors are installed. Hence, the entire building may be heated.

Fig. 5-11. Tarpaulins form a temporary enclosure when draped over timber framework. Sometimes a mudsill is used to support such framework.

Heating of Enclosures

For temporary enclosures the most economical and convenient source of heat will vary from area to area. Natural gas is often selected for its economy, but other heat sources such as electricity, steam, fuel oil, and bottled propane may be more available. Consideration should be given to the volume of air to be heated and the cost of bringing the source of heat to the construction site.

Heat loss or gain is calculated in terms of British thermal units (1 Btu = amount of heat required to raise the temperature of 1 lb. of water by 1 deg. F.). Natural gas rates are often expressed in therms (1 therm = 100,000 Btu), and electric rates are based on kilowatt-hours (1 kwh = 3,415 Btu). Portable heaters are classified by Btu of heat output. They are manufactured in various sizes ranging up to several million Btu.

Table 5-2 gives the heat loss of a tight polyethylene enclosure. It does not include heat loss through flaps (masons' entrances), cracks, and loose-fitting film. Thus, the tabulated values should be increased 25 to 50% if they are used as a guide to the size of heater required.

For example, suppose a polyethylene enclosure is 10 ft. wide, 50 ft. long, and 15 ft. high; a 30-mph wind is expected; and enclosure must be warmed 20 deg. F. Heat loss with a tight polyethylene enclosure would be 25 Btu per sq.ft. per hour, and assuming a 50% increase for other heat losses, total heat loss would be 86,000 Btu per hour. The required capacity of a heater must also include the Btu of heat necessary to put heat into the enclosure (3,600 Btu will heat 10,000 cu.ft. 20 deg. F.).

It may be possible to use the heating plant intended for the building if a release is signed for the heating contractor. For a large building it is usually necessary for the builder to supply a temporary heat source. Where only the wall under construction is enclosed, light portable heaters are applicable (Fig. 5-12).

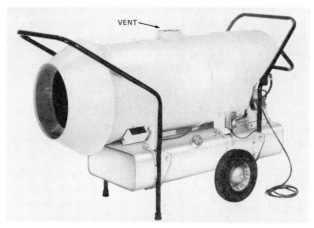

Fig. 5-12. A portable oil-fired space heater may be used to heat an enclosure. Note the provision for venting.

Although venting ductwork can be connected to some portable heaters, ideally a heater should be located outside the enclosure to blow hot air in. Venting of fossil-fuel burning heaters is very important because they produce carbon dioxide, which combines with calcium hydroxide in fresh mortar to form a weak layer of calcium carbonate or chalkiness on the surface. Venting such heaters also assures a greater supply of fresh air to the workmen, thus maintaining their health and productivity.

It should be noted that excessive *dry* heat will cause rapid drying of the mortar. Thus, live steam is an alternative heat source for enclosures, although it does have limitations; e.g., ice may form on the enclosure.

Other Construction Practices

One of the most important practices of concrete masonry work in subfreezing temperatures is to have the masonry units be delivered dry and kept dry until they are laid in the wall. In addition, they should be laid only on a sound, unfrozen surface and never on a snow- or ice-covered base or bed; not only is there danger of movement when the base thaws, but no bond will develop. It is also considered good practice to heat the surface of an existing masonry course to the same temperature as the masonry units to be added. The heat should be sustained long enough to thaw the surface thoroughly.

During cold weather the mortar should be mixed in smaller quantities than usual to avoid excessive cooling before use. Metal mortarboards with built-in electrical heaters may be used if care is taken to avoid overheating or drying the mortar. To avoid premature cooling of heated masonry units, only those units that will be used immediately should be removed from the heat source.

Table 5-2. Heat Loss of Polyethylene Film*

Wind velocity, mph	Heat loss, Btu per sq.ft. per hour		
	Per 1 deg. F.	Per 20 deg. F.	Per 50 deg. F.
0	0.70	14	35
5	0.93	19	46
10	1.05	21	52
20	1.17	23	58
30	1.23	25	61
40	1.26	25	63

*Source: Fig. 1, Chapter 20, of Ref. 29.

Regardless of temperature, concrete masonry units should never be wetted before being laid in a wall. In cold weather, wetting masonry units adds to the possibility of freezing any free water on the surfaces, increasing shrinkage, and defeating the goal of drawing off water from the mortar to a level below 6% (to prevent mortar expansion upon freezing).

Tooling time during cold weather is less critical than at normal temperatures. Hence, the mason may lay more masonry units before tooling. In all instances, however, the joint should be tooled before mortar hardens. Caution: premature tooling at day's end will cause lighter joints.

Fig. 5-13. Concrete masonry test prisms are covered with polyethylene and cured in a heated box.

Upon completion of each section or at the end of each workday, measures should be taken to protect new concrete masonry construction from the weather. During cold-weather construction, rapid drying out or early freezing of the mortar should be prevented. Furthermore, the top of the concrete masonry structure must be protected from rain or snow by plastic or canvas tarpaulins extending at least 2 ft. down all sides of the structure.

During cold-weather construction, careful attention to quality control should be exercised. As shown in Fig. 5-13, masonry test prisms should be covered and stored in a curing box in the same manner as concrete test cylinders are. The temperature inside the box should be maintained between 60 and 80 deg. F. (as required by ASTM C31, Standard Method of Making and Curing Concrete Test Specimens in the Field) or not less than 68 deg. F. (as required by the National Building Code of Canada).

Safety precautions require added emphasis during cold-weather construction for several reasons. Workmen not only direct more of their attention to comfort and out-of-the-ordinary construction problems due to low temperature, but also must overcome personal hazards such as uncertain footing on ice and snow and clumsiness due to protective clothing. Extra care must also be taken of flammable building materials and heaters since the possibility of fire and asphyxiation is increased.

In addition, enclosures for cold-weather protection should be securely guyed or braced if there is danger of snow and ice loads or wind forces. Snow and ice loads can contribute to overloading an *unheated* enclosure. Wind forces can make tarpaulins enclosing scaffolds act like sails. Also, the combination of wind forces and dead weight of a stack of masonry units can overload enclosure framework unless it has been carefully sized and installed.

Chapter 6

CONSTRUCTION TECHNIQUES FOR CONCRETE MASONRY

Masonry structures have been built for thousands of years and the methods passed on from generation to generation by instruction and example. The techniques of concrete masonry construction are well known to architects, engineers, and builders. Recommendations and explanations given in this chapter concern the current state-of-the-art.

Quantity Takeoffs

The first step in construction is to estimate the quantity of materials required. The tables on the next three pages may be used as guides for quantity takeoffs of concrete masonry materials.

Fig. 6-1. Tooling mortar joints of concrete split-block masonry. The fractured block produce walls of rugged appearance. Note untooled joints at left.

Table 6-1. Wall Weights and Material Quantities for Single-Wythe Concrete Masonry Construction*

Nominal wall thickness, in.	Nominal size (width x height x length) of concrete masonry units, in.	Average weight of 100-sq.ft. wall area, lb.**		Material quantities for 100-sq.ft. wall area		Mortar for 100 units, cu.ft.††
		Units made with sand-gravel aggregate†	Units made with lightweight aggregate†	Number of units	Mortar, cu.ft.	
4	4x4x16	4,550	3,550	225	13.5	6.0
6	6x4x16	5,100	3,900	225	13.5	6.0
8	8x4x16	6,000	4,450	225	13.5	6.0
4	4x8x16	4,050	3,000	112.5	8.5	7.5
6	6x8x16	4,600	3,350	112.5	8.5	7.5
8	8x8x16	5,550	3,950	112.5	8.5	7.5
12	12x8x16	7,550	5,200	112.5	8.5	7.5

*Based on 3/8-in. mortar joints.
**Actual weight of 100 sq.ft. of wall can be computed by the formula $WN + 145M$ where W is the actual weight of a single unit; N, the number of units for 100 sq.ft. of wall; and M, the mortar (cu.ft.) for 100 sq.ft. of wall.
†Based on Fig. 1-3 (page 7), using a concrete density of 138 pcf for units made with sand-gravel aggregate and 87 pcf for lightweight-aggregate units, and the average weight of unit for two- and three-core block.
††With face-shell mortar bedding. Mortar quantities include allowance for waste.

For single-wythe walls, wall weights and material quantities are given in Table 6-1. Table 6-2 lists material quantities for composite walls bonded with metal ties or masonry headers. The mortar quantities include the customary allowance for waste that occurs during construction for a variety of reasons.

The breakdown of materials in 1 cu.ft. of mortar is given in Table 6-3. Note that some of the sample mixes given fit the classification of two types of mortar. This is because mortar specifications permit a range in the amounts of lime and sand used for any type of mortar. For all practical purposes, the limits of the range of proportions for the various types of mortar coincide.

Tables 6-4 and 6-5 give grout quantities for grouted walls of concrete brick or concrete block. These tables are also useful for estimating grout in reinforced concrete masonry.

Construction Procedures

Keeping Units Dry

When delivered to the job, concrete masonry units should be dry enough to comply with specified limitations for moisture content. To be maintained in this dry condition, they should be stockpiled on planks or other supports free from contact with the ground and then be covered with roofing paper or canvas or polyethylene tarpaulins. The top of a concrete masonry structure should be covered with tarpaulins or boards to prevent rain or snow from entering unit cores during construction.*

Concrete masonry units should never be wetted immediately before and during placement, a practice customary with some other masonry materials. As discussed in Chapter 4, when moist concrete masonry units are placed in a wall, they will shrink with the loss of moisture. If this shrinkage is restrained, as it usually is, stresses develop that may cause cracks in the walls. Hence, it is important that the units be kept at or dried to at least the moisture content limitations of the applicable specifications.

Sometimes it may be advisable to dry concrete masonry units below the moisture content usually specified for the locality; e.g., where walls will be exposed to relatively high temperatures and low humidities in interiors of heated buildings. In such cases it is advisable that, before placement, the units be dried to approximately the average air-dry condition to which the finished construction will be exposed in service.

Damp concrete masonry units can be stacked to facilitate drying and then artificially dried by blowing heated air through the cores and the spaces between the stacked units. An inexpensive and efficient drying device consists of a combination oil- or gas-burning heater and fan. This method of drying works equally well indoors or outdoors and can readily be used in the plant or at the jobsite.

---

*See Chapter 5 for hot- and cold-weather construction practices, including coverings and enclosures.

Table 6-2. Material Quantities (Concrete Block, Brick, and Mortar) for 100-Sq.Ft. Area of Composite Walls

Wall thickness, in.	Type of bonding*	No. and size, in., of block		No. of brick	Mortar, cu.ft.**
		Stretchers	Headers		
8	A—metal ties	112.5— 4x8x16	—	675	20.0
	B—7th-course headers	97— 4x8x16	—	770	12.2
	C—7th-course headers	197— 4x5x12	—	770	13.1
12	D—metal ties	112.5— 8x8x16	—	675	20.0
	E—7th-course headers	97— 8x8x16	—	868	13.5
	F—6th-course headers	57— 8x8x16	57— 8x8x16	788	13.6

*Key for type shown below.
**Mortar quantities based on 3/8-in. mortar joints with face-shell bedding for the block; mortar quantities include allowance for waste. All unit sizes are nominal.

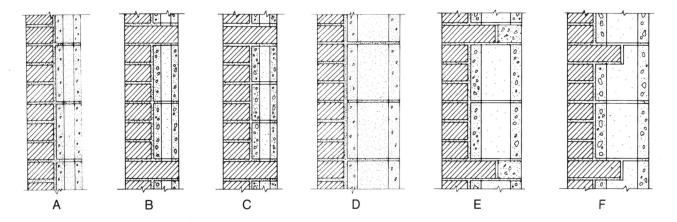

A   B   C   D   E   F

Table 6-3. Sample Quantities of Mortar Materials

Mortar type*	Mix proportions, parts by volume**				Material quantities, cu.ft., for 1 cu.ft. of mortar			
	Portland cement	Masonry cement	Hydrated lime	Sand	Portland cement	Masonry cement	Hydrated lime	Sand
M	1	1	—	6	0.16	0.16	—	0.97
M or S	1	—	1/4	3	0.29	—	0.07	0.96
N	1	—	1	6	0.16	—	0.16	0.97
N or O	—	1	—	3	—	0.33	—	0.99

*See Chapter 2, "Specifications and Types," for U.S. or Canadian mortar mix ranges by types (page 32).
**The proportions shown here fall within the range of mixes allowed. See Table 2-1 (page 31) for proportion specifications for mortar.

Table 6-4. Volume of Grout in Two-Wythe Grouted Concrete Brick Walls*

Width of grout space, in.	Grout, cu.yd., for 100-sq.ft. wall area**	Wall area, sq.ft., for 1 cu.yd. of grout**
2.0	0.64	154
2.5	0.79	126
3.0	0.96	105
3.5	1.11	89
4.0	1.27	79
4.5	1.43	70
5.0	1.59	63
5.5	1.75	57
6.0	1.91	53
6.5	2.06	49
7.0	2.22	45

*Adapted from *Volume of Grout Required in Masonry Walls*, Design Aid 15, Masonry Institute of America, Los Angeles, Calif., 1971.
**A 3% allowance has been included for waste and job conditions.

Table 6-5. Volume of Grout in Grouted Concrete Block Walls*

Wall thickness, in.	Spacing of grouted cores, in.	Grout, cu.yd., for 100-sq.ft. wall area**	Wall area, sq.ft., for 1 cu.yd. of grout**
6	All cores grouted	0.79	126
	16	0.40	250
	24	0.28	357
	32	0.22	450
	40	0.19	526
	48	0.17	588
8	All cores grouted	1.26	79
	16	0.74	135
	24	0.58	173
	32	0.49	204
	40	0.44	228
	48	0.39	257
12	All cores grouted	1.99	50
	16	1.18	85
	24	0.91	110
	32	0.76	132
	40	0.70	143
	48	0.64	156

*Adapted from *Volume of Grout Required in Masonry Walls*, Design Aid 15, Masonry Institute of America, Los Angeles, Calif., 1971.
**A 3% allowance has been included for waste and job conditions. All quantities include grout for intermediate and top bond beams in addition to grout for cores.

Mortaring Joints

Two types of mortar bedding are used with concrete masonry: full mortar bedding and face-shell mortar bedding (Fig. 6-2). In full mortar bedding, the unit webs as well as face shells are bedded in mortar. Full bedding is used for laying the first or starting course of block on a footing or foundation wall as well as for laying solid units such as concrete brick and solid block. It is also commonly used when building concrete masonry columns, piers, and pilasters that will carry heavy loads. Where some vertical cores are to be solidly grouted, such as in reinforced masonry, the webs around each grouted core are fully mortared. *For all other concrete masonry work with hollow units, it is common practice to use only face-shell bedding.* Also, the head (vertical) joints of block having plain ends are mortared only opposite the face shells.

Block

For bed (horizontal) joints, all concrete block should be laid with the thicker part of the face shell up. This provides a larger mortar-bedding area and makes the block easier to lift. A mechanical mortar spreader (Fig.

6-3) can be used to speed production, especially on long, straight walls.

For head joints, mortar is applied only on the face-shell ends of block. Some masons butter (mortar) the vertical ends of the block previously placed; others set the block on one end and butter the other end before laying the block. Time can be saved by placing three or four block on end and then buttering their vertical edges in one operation (Fig. 6-4). Some masons butter both the block already laid and the block to be laid (Fig. 6-5); such application of mortar ensures well-filled head joints.

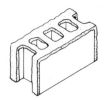
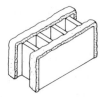

Full mortar bedding   Face-shell mortar bedding

Fig. 6-2. Types of mortar bedding. Only face-shell mortar bedding is ordinarily used to lay concrete block.

Fig. 6-3. A mechanical mortar spreader equipped with an electric vibrator applies mortar to the face shells of a block wall.

Fig. 6-4. Most masons spread mortar on the bed joint for only three block at a time and then butter the head joints on them so all three can be placed in rapid succession. This is convenient because usually the three block plus mortar joints equal 4 ft., the same length as a mason's level.

Fig. 6-5. A well-filled head joint results from mortaring both block.

Fig. 6-6. Each block is pushed downwards and sideways so that mortar oozes out of the head and bed joints.

Regardless of the method used to apply mortar to the vertical edges, each block is brought over its final position and pushed downward into the mortar bed and sideways against the previously laid block so that mortar oozes out of the head and bed joints on both sides of the face shell (Fig. 6-6). This indicates that the joints are well filled.

Caution: Mortar should not be spread too far ahead of the actual laying of units or it will tend to stiffen and lose its plasticity, thereby resulting in poor bond. In hot, dry weather it may be necessary to spread only enough mortar for each block as it is laid.

Fig. 6-7. For laying brick, the mason spreads mortar uniformly on the bed joints. Even slight furrowing, as shown here, is sometimes not permitted.

Fig. 6-8. Brick is pressed into place so that mortar oozes out of the head and bed joints.

As each block is laid, excess mortar extruding from the joints is cut off with the trowel and thrown back on the mortarboard for reuse. Some masons apply the extruded mortar to the vertical face shells of the block just laid. If there has been any delay long enough for the extruded mortar to stiffen on the block before it is cut off, it should be reworked on the mortarboard before reuse. Mortar droppings picked up from the scaffold or from the floor should not be reused.

Brick

For concrete brick as well as clay brick, mortar should be spread uniformly thick for the bed joints and furrowed only slightly *if at all* (Fig. 6-7). Many specifications and some building codes prohibit furrowing on bed joints. The weight of the brick and the courses above help compact the mortar and ensure watertight bed joints.

In brick construction, special care should be taken in filling the head joints because they are more vulnerable to water penetration than the bed joints. If head joints are not completely filled with mortar, some voids and channels may permit water to penetrate to the inside of the wall. Plenty of mortar should be troweled on the end of the brick to be placed so that, when the brick is shoved into place, the mortar will ooze out at the top and around the sides of the head joint, indicating the joint is completely filled (Fig. 6-8). Dabs of mortar spotted on both corners of the brick do not completely fill the head joints, and "slushing" (attempting to fill the joints from above after the brick is placed) cannot be relied on to fill all voids left in the head joints.

Laying Up a Wall

First Course

On jobs where more than one mason is working, the footing or slab foundation must be level so that each mason can start his section of wall on a common plane and the bed joints will be uniformly straight when the sections are connected. If the foundation is badly out of level, the entire first course should be laid before masons begin work on other courses, or a level plane should be established with a transit or engineer's level.

After the corners are located, masons often string out the masonry units for the first course without mortar in order to check the wall layout (Fig. 6-9). A chalk line is sometimes used to mark the foundation and help align the block accurately.

Before any units are laid, the top surface of the concrete foundation should be clean. Laitance is removed and aggregate exposed by sandblasting, chipping, or scarifying if necessary to ensure a good bond of the masonry to the foundation. Then a full mortar bed is spread and furrowed with a trowel to ensure plenty of mortar along the bottom edges of the block for the first course (Fig. 6-10). If the wall is to be grouted, the mortar bedding for the first course should *not* fill the area under the block cores; grout should come into direct contact with the foundation.

The corner block should be laid first and carefully positioned. After three or four block have been laid, the mason's level is used as a straightedge to assure correct alignment of the block (Fig. 6-11). Then these units are carefully checked with the level and brought to proper grade (Fig. 6-12) and made plumb (Fig. 6-13) by tapping with the trowel handle. The entire first course of a concrete masonry wall should be laid with such care, making sure each unit is properly aligned, leveled, and plumbed. This will assist masons in laying succeeding courses. Any error at this stage—in the first course—means continuing trouble in laying up a straight, true wall.

Fig. 6-11. Corners are laid first with great care. The mason's level is used as a straightedge to assure correct alignment of the block.

Fig. 6-9. To check the layout, the mason may string out the block without mortar.

Fig. 6-12. Block are leveled by tapping with the trowel handle.

Fig. 6-10. A full mortar bed is necessary for laying the first course.

Fig. 6-13. The trowel handle is also used to make the block plumb (vertically straight).

Corners

The corners of the wall are built up higher, usually four or five courses higher, than the course being laid at the center of the wall. As each course is laid at a corner, it is checked with a level for alignment (Fig. 6-14), for being level (Fig. 6-15), and for being plumb (Fig. 6-16). In addition, each block is carefully checked with a level or straightedge to make certain that the faces of all the block are in the same plane (Fig. 6-17).

Other precautions are necessary at corners to ensure true, straight walls. An accurate method of finding the top of the masonry for each course is provided by the use of a simple story pole made from a 1×2-in. wood strip with markings 8 in. apart (Fig. 6-18); more elaborate metal story poles are commercially available. Also, since each course is stepped back a half unit, the mason easily checks the horizontal spacing of the units by placing his level diagonally across their corners (Fig. 6-19).

Between Corners

After the corners at each end of a wall have been laid up, a mason's line (string line) is stretched tightly from corner to corner for each course and the top outside edge of each block is laid to this line (Fig. 6-20). The line is moved up as each course is laid.

The manner of handling and gripping a masonry unit is important, and the most practical way for each individual is determined through practice. Generally, the mason tips a block slightly towards himself so that he can see the upper edge of the course below and thus place the lower edge of the block directly over it. By rolling the block slightly to a vertical position and shoving it against the previously laid unit, he can lay the block to the mason's line with minimum adjustment. This speeds the work and reduces the possibility of breaking mortar bond—by not moving the unit excessively after it has been pressed into the mortar. Light tapping with the trowel handle should be the only adjustment necessary to level and align the unit to the mason's line (Fig. 6-21). The use of the mason's level between corners is limited to checking the face of each unit to keep it aligned in a true plane with the face of the wall.

Caution: All adjustments to final position must be made while the mortar is soft or plastic. Any adjustments made after the mortar has stiffened or even partially stiffened will break the mortar bond and cause cracks between the masonry unit and the mortar. This would allow penetration of water. Any unit disturbed after the mortar has stiffened should be removed and relaid with fresh mortar. Realignment of a unit should not be attempted after a higher course has been laid.

Care must be taken to keep the wall surface clean during construction. In removing excess mortar that has oozed out at the joints, the mason must avoid

Fig. 6-14. When laying up corners, the mason checks each course for alignment.

Fig. 6-15. Checking corner for level.

Fig. 6-16. Checking corner for plumb.

Fig. 6-17. Checking each block at a corner to see that it is in plane.

Fig. 6-18. Using a story pole to check the spacing of courses.

Fig. 6-19. Diagonal check of the horizontal spacing of units.

Fig. 6-20. With string line in place, the mason lays block between corners. Masons have individual ways of gripping the block.

Fig. 6-21. Light tapping brings the block into position with the string line.

Fig. 6-22. To prevent smears, mortar droppings should be almost dry when cut off the wall with a trowel.

Fig. 6-24. Brushing—the final step in mortar removal.

Fig. 6-23. Most of the dry mortar remaining on the wall can be rubbed off with a piece of concrete masonry.

smearing soft mortar onto the face of the unit, especially if the wall is to be left exposed or painted. Numerous embedded mortar smears will detract from the neat appearance of the finished wall. They can never be removed and paint cannot be depended on to hide them.

Any mortar droppings that do stick to the wall should be almost dry before they are removed with a trowel (Fig. 6-22). Then, when dry and hard, most of the remaining mortar can be removed by rubbing with a small piece of concrete masonry (Fig. 6-23) and by brushing (Fig. 6-24).

Closure Unit

Before the closure unit is laid in a course, the length of the opening is checked in order to avoid joints that are too tight or too wide. If necessary, the closure unit

is accurately measured, sawed, and dressed for a proper fit in the opening.

When installing the closure unit, the mason butters all edges of the opening and all four vertical edges of the closure unit (Fig. 6-25) before he carefully lowers the unit into place (Fig. 6-26). If any of the mortar falls out, leaving an open joint, the mason removes the closure unit, applies fresh mortar, and repeats the operation. Closure unit locations are staggered throughout the length of the wall.

Building Composite Walls

In composite concrete masonry walls the first wythe laid is parged (backplastered) with mortar not less than 3/8 in. thick before the adjacent wythe is laid (Fig. 6-27). This provides composite action of the two wythes and an effective barrier against water penetration through the wall. Before parging, however, mortar protruding from the joints of the first-laid wythe should be cut flush while the mortar is still soft; otherwise, parging over the hardened mortar may break the bond in the mortar joints and result in a leaky wall.

If facing is laid first, a mason's level is often used during parging to support the facing and prevent it from tilting and breaking bond in the mortar joints (Fig. 6-28). With a light facing wythe, however, it is best to lay up the block backing first so that the parging can be applied on the more substantial wythe (Fig. 6-29).

In composite construction, all solid facing units should be laid with *full* mortar bedding and the head joints completely filled. In header courses the cross joints also should be completely filled; i.e., mortar is spread over the entire side of the header unit before it is shoved into place.

Fig. 6-25. All edges of the opening for the closure block are buttered.

Fig. 6-28. A level is often used during parging to brace the facing.

Fig. 6-26. The closure block is carefully lowered into place.

Fig. 6-29. The block backup is the more substantial wythe on which to apply parging.

Fig. 6-27. The first wythe is parged before the adjacent wythe is laid.

Fig. 6-30. Header block can be laid with the notch up.

Fig. 6-32. Concrete brick can be used to back up the brick headers.

Fig. 6-31. Here the notch of the header block is down.

Fig. 6-33. Cavity wall.

Fig. 6-34. To keep the cavity clean, a wood strip is laid across the ties in the cavity.

Fig. 6-35. The wood strip is lifted to remove mortar droppings.

For walls greater than 10 in. in thickness, specially shaped header block are sometimes used to bond the facing headers and the backup units with 6th-course bonding, as shown previously in Table 6-2. The special header block can be laid with the recessed notch up (Fig. 6-30) or down (Fig. 6-31), and with block backup laid in face-shell mortar bedding. In 7th-course bonding, the backup consists of stretcher block only, and concrete brick may be used as backup to the brick headers (Fig. 6-32).

Building Cavity Walls

Cavity walls generally consist of two single-wythe walls that are separated by a continuous air space 2 to 4 in. wide but connected by rigid metal ties embedded in the mortar joints (Fig. 6-33). The ties should have sufficient mortar cover, as shown in Fig. 4-5c (page 73).

During construction of the wythes, the cavity must be kept free of mortar droppings that could form a bridge for moisture to pass across the cavity to the interior wythe. A simple method of maintaining a clear cavity is to catch mortar droppings on a board laid across a tier of ties (Fig. 6-34). When the masonry reaches the next level for ties, the board is raised, cleaned, and repositioned (Fig. 6-35).

In addition, dropping mortar into the cavity may be avoided by spreading the mortar bed so that it is back about 1/2 in. from the edge on the cavity side. When the next masonry units are laid, the mortar spreads to the edge of the unit without squeezing out into the cavity. Another method commonly used is to spread the mortar and then draw the trowel over the mortar in an upward and outward direction away from the cavity, thus forming a beveled mortar bed. When the units are laid on such a beveled bed, the mortar spreads only to the cavity edge.

Any mortar fins occurring on the inner faces of the cavity should be troweled flat. This not only prevents the mortar from falling into the cavity but also provides a smooth surface to facilitate placement of granular insulating materials if required later.

During installation of flashing* in cavity walls, care must be taken to avoid tearing or puncturing the flashing material, thus destroying its effectiveness. Through-wall flashing should be laid over a thin bed of mortar and then another thin layer of mortar placed on top of the flashing to act as bedding for the next masonry course. The seams of the flashing should be thoroughly bonded to ensure continuity of the flashing and prevent penetration of water. Most sheet metal flashing can be soldered, but it requires lockslip joints at intervals to permit thermal expansion and contraction. Plastic flashing, generally joined by heat or appropriate adhesives, does not require expansion seams because it is usually elastic enough to take a certain amount of stretching.

---

*Additional information on flashing as well as weephole construction is given at the end of Chapter 4.

Building Reinforced Walls

For reinforced masonry wall construction the procedures used in laying masonry units, placing reinforcing bars, and pouring grout vary with the size of the job, the equipment available, and the preferences of the contractor. Therefore, this section covers only the general requirements of common procedures.

Procedures Before Grouting

Solid or hollow concrete masonry units should be laid so that their alignment forms an unobstructed, continuous series of vertical cores within the wall framework. Spaces in which reinforcement will be placed should be at least 2 in. wide. No grout space should be less than 3/4 in. or more than 6 in. wide; if the grout space is wider than 6 in., the wall section should be designed as a reinforced concrete member.

Two-core, plain-end units are preferable to three-core units because the larger cores allow easier placement of reinforcing bars and grout. Also, these units are more easily aligned for their cores to form continuous, vertical spaces. When A-shaped, open-end masonry units (Fig. 4-9, page 76) are used, they are arranged so the closed ends are not abutting.

Many building codes specify Type PL mortar for reinforced concrete masonry; however, Type PM mortar may be used if desired. The mortar bed under the first course of block should not fill the core area because the grout must come into direct contact with the foundation. All head and bed joints should be filled solidly with mortar for the thickness of the face shell. With plain-end units, however, it is not necessary to fill the head joint across the full unit width. Also, when the wall is to be grouted intermittently (for reinforcement 16, 24, 32, or 48 in. on center), only the webs at the extremity of those cores containing grout are mortared. When the wall is to be solidly grouted, none of the cross webs need be mortared since it is desirable for the grout to flow laterally and form the bed joints at all web openings.

Mortar protrusions that cause bridging and thus restrict the flow of grout require an excessive amount of vibration or puddling to assure complete filling of the grout space. Hence, care is necessary that mortar projecting more than 3/8 in. into the grout space be removed and that excess mortar does not extrude and fall into the grout space. The mason can prevent mortar from extruding into the grout space by placing the mortar no closer than 1/4 to 1/2 in. from the edge of the grout space and troweling the mortar bed upward and outward, away from the edge, thus forming a bevel. Mortar droppings in the grout spaces of multiwythe walls can be caught and removed by using a wood strip as described in the preceding section on cavity wall construction.

Vertical reinforcement may be erected before or after the masonry units are laid. When the reinforcing bars are placed before the units, the use of two-core, open-end, A- or H-shaped units (Figs. 4-9 and 1-22c, pages 76 and 21, respectively) becomes desirable in order for the units to be threaded around the reinforcing steel. When the bars are placed after the units, adequate positioning devices are required to prevent displacement during grouting. Both vertical and horizontal reinforcement should be accurately positioned and rigidly secured at intervals by wire ties or spacing devices (Fig. 6-37).

Horizontal reinforcement is placed as the wall rises. The reinforcing bars are positioned in bond-beam, lintel, or channel units, which are then solidly grouted (Fig. 6-38). Where the wall itself is not to be solidly grouted and cored bond-beam units are used, the grout may be contained over open cores by placing expanded metal lath in the horizontal bed joint before the mortar bed is spread for the bond-beam units. Paper or

Fig. 6-36. Reinforced concrete masonry walls are used in earthquake zones.

wood should not be used as a grout barrier because of fire resistance requirements.

To ensure solid grouting of bond beams, it may be necessary to fill those portions of the bond beams between the vertically grouted cores as the bond-beam courses are laid, especially if the spacing of vertically grouted cores is greater than 4 ft. Otherwise, the grout may not flow far enough horizontally from the cores being grouted to completely fill the bond beams.

A concrete masonry wall should be grouted as soon as possible to reduce shrinkage cracking of the joints. However, placing grout before the mortar has been allowed to cure and gain strength may cause shifting or blowout of the masonry units during the grouting operations. Therefore, to fill the cavity in two-wythe masonry or in large cavities of masonry sections (made up of two or more units and containing vertical joints, such as pilaster sections), grout should be poured after the mortar in the entire height of the

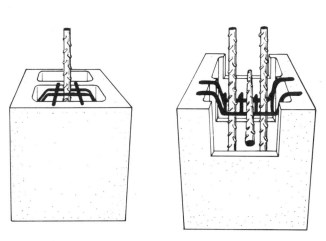

Fig. 6-37. Reinforcing bar spacers.

masonry has been cured a minimum of 3 days during normal weather or 5 days during cold weather. The hydrostatic or fluid pressure exerted by freshly placed grout on the masonry shell may be ignored when filling hollow-core masonry units, and thus it is unnecessary to cure mortar in hollow-unit masonry walls for longer than 24 hours before grouting.

Low-Lift Grouting

Of the two grouting procedures in general use—low- and high-lift* grouting—low-lift grouting is the simplest and most common. This procedure requires no special concrete masonry units or equipment.

In low-lift grouting of a single-wythe wall, the wall is built to a height not exceeding 5 ft. before grout is pumped or poured into the cores. This operation is repeated by alternately laying units and grouting at successive heights not exceeding 5 ft. In high-lift grouting, the wall is built to full story-height first before grouting the cores or cavities.

Typical reinforced, single-wythe, hollow masonry construction using low-lift grouting is shown in Fig. 6-38. Vertical cores to be filled should have an unobstructed alignment, with a minimum dimension of 2 in. and a minimum area of 8 sq.in. Also, the vertical reinforcing bars may be relatively short in length because they only need to extend above the top of the lift a distance equal to 30 bar diameters for sufficient overlap with the reinforcing bars in the next lift. As an alternate choice, vertical steel may extend to full wall height for one-story construction or to ceiling height (plus overlap) for multistory construction. However, since the long lengths of steel require the use of

---

*A "lift" is the layer of grout pumped or poured in a single continuous operation. A "pour" is considered to be the entire height of grouting completed in one day; it may be composed of a number of successively placed grout lifts.

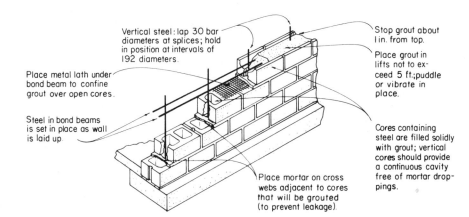

Fig. 6-38. Low-lift grouting of a typical single-wythe reinforced masonry wall.

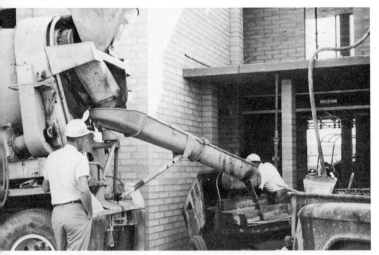

Fig. 6-39. Ready mixed portland cement grout is delivered by truck mixer into the pump hopper. Note the hose extending to work above.

Fig. 6-40. While helping to handle the grout hose, a laborer controls the pump shutoff with a hand button.

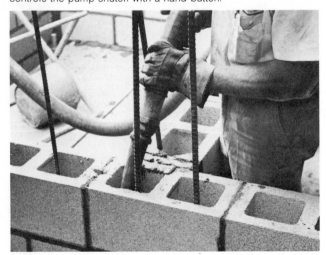

Fig. 6-41. View of grout shooting from the hose into a core.

open-end units, some masonry contractors prefer to lap the steel just above each 5-ft. lift.

Grout is handled from the mixer to the point of deposit in the grout space as rapidly as practical (Figs. 6-39, 6-40, and 6-41). Pumping or other placing methods that prevent segregation of the mix and limit grout splatter are used. On small projects, the grout is poured with buckets having spouts or funnels to confine the grout and prevent splashing or spilling onto the face or top surface of the masonry. Grouting should be done from the inside face of the wall if the outside will be exposed; dried grout can deface the exposed surface of a wall and be detrimental to the mortar bond of the next masonry course. On most projects, grout pumps are recommended to save time and money.

Whenever work is stopped for one hour or longer, a horizontal construction joint should be made by stopping the grout pour about 1 in. below the top of the masonry unit to form a key with the next lift (Fig. 6-42). This is required by most codes.

During placement, grout should be rodded (usually with a 1×2-in. wooden stick) or mechanically vibrated to ensure complete filling of the grout space and solid embedment of the reinforcement. It takes very little effort to do this consolidation job properly because of the fluid consistency of the grout. When high-absorption masonry units are used, it may be necessary to rerod or revibrate the grout 15 to 20 minutes after placement to overcome any separations (of the grout from the reinforcing steel) and voids caused by settlement of the grout and absorption of water into the surrounding masonry. Over-vibration, however, must be carefully avoided at this stage; more hazardous than during initial consolidation, it can cause blowouts, broken ties, cracked masonry units, or segregation of the grout.

Fig. 6-42. Grouting is stopped about 1 in. below the top of the block to form a key with the next lift.

In low-lift grouting of masonry with two or more wythes, the exterior wythe is laid up a maximum of 18 in. above the interior wythe. After the interior wythe is laid, the cavity between the wythes is grouted in lifts not to exceed six times the width of the grout space, with a maximum of 8 in. No minimum mortar curing period is necessary before grouting. Grout is poured into the grout space to within 1 in. of the top of the interior wythe and then consolidated.

Where there are more than two wythes, the middle wythe (usually of brick size) may be built by "floating" the units in the grout space; i.e., pushing the units down into the grout so that a 3/4-in. depth of grout surrounds the sides and ends of each unit. No units or pieces of a unit less than 10 cu.in. in size should be embedded in the grout by floating.

High-Lift Grouting

With this procedure, grouting is delayed until the wall has been laid up to full story-height. High-lift grouting is intended for use on wall construction where reinforcement, openings, or masonry unit arrangements do not prevent the free flow of grout or inhibit the use of mechanical vibration to properly consolidate the grout in all cores or horizontal grout spaces. The vertical cores should have an unobstructed alignment, with a minimum dimension of 3 in. and a minimum area of 10 sq.in. In two-wythe masonry the minimum dimension of the grout space (cavity) between wythes is 2 or 3 in., depending on the governing code, and the maximum is 6 in.

Vertical bulkheads extending the entire height of the wall should be provided at about 26 ft. on center to control the flow of the grout horizontally. In a hollow-unit masonry wall, such barriers are made by placing mortar on cross webs and blocking the bond-beam units with masonry bats set in mortar. In a multi-wythe wall, the barriers are laid into the grout spaces as the wall is erected. In addition to confining grout to a manageable area, these barriers may be used as stiffeners or points to locate wall bracing.

Proper preparation of grout space is one of the most important features of high-lift grouting. It is necessary, before grouting, to remove all mortar droppings and debris through cleanout openings. Not less than 3×4 in. in size, a cleanout opening is located at the bottom of every core in hollow-unit reinforced masonry containing dowels or vertical reinforcement, and in at least every other core that is grouted but has no vertical bars. In solidly grouted plain masonry, cleanouts should be provided by leaving out every other unit in the bottom tier. In a two-wythe masonry wall, the cleanouts are provided at the bottom of the wall by omitting alternate units in the first course of one wythe. The governing standard or building code should be consulted to verify requirements for cleanout openings.

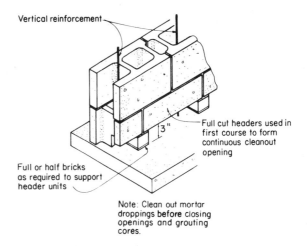

Fig. 6-43. Cleanout opening detail (alternate).

Cleanout openings in the face shells of units should be made before the units are laid. A special scored unit that permits easy removal of part of a face shell is occasionally used. Also, an alternate cleanout design makes use of header units, as shown in Fig. 6-43.

It is considered good practice to cover the bottom of a grout space with a 2- to 3-in. layer of sand or a polyethylene sheet to act as a bond-breaker for the mortar droppings. The grout space is flushed at least twice a day (at midday and quitting time) with a high-pressure stream of water or, to keep the masonry from being moistened unnecessarily, the mortar droppings and projections are dislodged with a long pole or rod as the work progresses. After the masonry units are laid, the sand or polyethylene sheet is removed; compressed air is used to blow any remaining mortar out of the grout space; and the space is checked for cleanliness and the reinforcement for position. A mirror is a good inspection tool for looking up into the grout space through a cleanout opening.

Before grouting, the cleanout openings are closed by inserting masonry units or the face shells that were left out, or by placing formwork over the openings to allow grouting right up to the wall face. Grouting need not be delayed until the face-shell plugs or cleanout closure units are cured, but they should be adequately braced to resist the grout pressure.

In high-lift grouting, intermediate horizontal construction joints are usually not permitted. Once the grouting of a wall section is started, one pour* of grout to the top of the wall (generally in 5-ft. maximum lifts) should be planned for a workday. Should a blowout, an equipment breakdown, or any other emergency stop the grouting operation, a construction joint may be used if approved by the inspector. The alternatives

---

*See footnote on page 133.

are to wash out the fresh grout or else rebuild the wall.

For economical placement a uniform 5-ft. lift of grout is generally pumped into place and immediately (not more than 10 minutes later) rodded or, preferably, vibrated. Each succeeding lift of grout is pumped and consolidated after an appropriate lapse of time, a minimum of 30 and a maximum of 60 minutes (depending upon weather conditions and masonry absorption rates), to allow for settlement shrinkage and the absorption of excess water by the masonry units. This waiting period also reduces the hydrostatic pressure of the grout and thus the possibility of blowout. In each lift, the top 12 to 18 in. of grout is reconsolidated before or during placement of the succeeding lift. For solid-grouted, hollow-unit masonry, only alternate-core grout need be reconsolidated.

In multi-wythe construction, the total length of wall that can be grouted in one pour is limited. It is determined by the number of sections (bounded by vertical bulkheads) that can be grouted to maintain the one-hour maximum interval between successive lifts in any section.

The maximum height of a pour is limited by practical considerations, such as segregation of grout, the effect of dry grout deposits left on the masonry units and the reinforcing steel, and the ability to consolidate the grout effectively. Under some circumstances, the maximum height of pour may be 12 ft. for 8-in., single-wythe, hollow-unit walls and 16 ft. for similar 12-in. walls; in multi-wythe construction, the maximum height of pour may be 16 ft. for walls with a single curtain of reinforcing steel (walls less than 12 in. thick) and 20 ft. for walls with two curtains of reinforcement (walls 12 in. or more in thickness). On the other hand, the height of pour may be governed by story height, and thus 8-in., single-wythe walls may have a 20-ft. height of pour. When the grout pour exceeds 8 ft. in height, building codes sometimes require special inspection of the work.

Extreme care should be used to prevent grout from staining any masonry wall that will be exposed to view. If grout does contact the face of the masonry, it should be removed immediately. Also, soon after the wall has been fully grouted, all exposed faces showing grout scum or stains (percolated through the masonry and joints) should be washed down thoroughly with a high-pressure stream of water. If necessary, further cleaning may be done after curing and before final acceptance by the architect.

The time- and money-saving advantages of high-lift grouting on large projects are obvious. The vertical steel can be placed after the wall is erected and, even on a job of moderate size, the grout can be supplied by a ready mixed concrete producer and pumped in a continuous operation. The main disadvantages of high-lift grouting may be the need for a grout pump or other means of pouring grout rapidly, and the requirement for cleanout openings at the base of the wall.

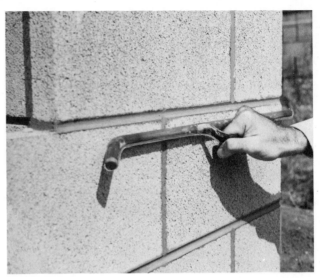

Fig. 6-44. Tooling a horizontal joint (concave type).

Tooling Mortar Joints

Weathertight joints and the neat appearance of concrete masonry walls are dependent on proper tooling; i.e., compressing and shaping the mortar face of a joint with a special tool slightly larger than the joint. After a wall section has been laid and the mortar has become thumbprint-hard (when a clear thumbprint can be impressed and the cement paste does not adhere to the thumb), the mortar joints are usually considered ready for tooling. Upon hardening, mortar has a tendency to shrink slightly and pull away from the edges of the masonry units. Proper use of a jointing tool restores the intimate contact between the mortar and the units and helps to make weathertight joints—by sealing any cracks left between the mortar and the units when they were laid. Proper tooling also produces uniform joints with sharp, clean lines.

Horizontal joints should be tooled before vertical joints. A jointer for tooling horizontal joints should be at least 22 in. long, preferably longer, and upturned on at least one end to prevent gouging (Fig. 6-44), while a jointer for vertical joints is small and S-shaped (Fig. 6-45). Plexiglass jointers are available to avoid staining white or light-colored mortar joints. After the joints have been tooled, any mortar burrs should be trimmed off flush with the face of the wall by using a trowel (Fig. 6-46) or by rubbing with a burlap bag, brush, or carpet.

The principal types of mortar joints used in concrete masonry are shown in Fig. 6-47. The concave, vee, raked, and beaded types need special jointing tools, whereas the flush, struck, and weathered types are finished with a trowel. The extruded (also called skintled or weeping) type is made by using extra

Fig. 6-45. Tooling a vertical joint.

Fig. 6-46. Mortar burrs are trimmed off after the joints are tooled.

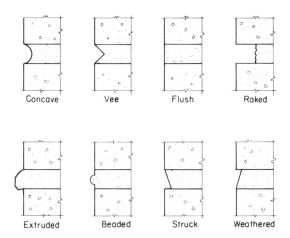

Fig. 6-47. Principal types of mortar joints.

mortar so it can be squeezed out or extruded as the units are laid; it is not trimmed off but left to harden. Extruded joints are *not* recommended for walls subject to heavy rains, high winds, or freezing temperatures.

Vee joints are usually narrow in appearance and have sharp shadow lines. Concave joints (pictured in Fig. 6-44) have less pronounced shadows. Both types are tooled and very effective in resisting rain penetration. They are recommended for exterior weathertight walls, as is the weathered type.

Flush joints are simple to make because the excess mortar is simply trimmed off with a trowel (striking downward rather than upward) and the flush face rubbed with a carpet-covered wood float. The mortar is not compacted by tooling, and small hairline cracks produced when the mortar is pulled away from the units by the trimming action may permit infiltration of water into the wall. Flush joints are used in walls that will be plastered.

Raked joints (pictured in Fig. 6-48) are made with a special tool—a joint raker or skate—to remove the mortar to a certain depth, which should not be more than 1/2 in. These joints produce dark shadows that accent the masonry pattern. Since their ledges may hold rain, snow, or ice that may affect the watertightness of the wall, they are best suited to dry climates or interior use.

Beaded joints are basically extruded joints that are tooled with a special bead-forming jointer. The beads protruding on the wall surface present strong shadow lines, but special care is required to obtain a straight appearance.

The struck and weathered types of joints are generally used to emphasize horizontal lines. Struck joints

137

are easy to make with a trowel, especially if the mason works from the inside of the wall. However, their small ledges do not shed water readily, making them unsuitable for use in areas where heavy rains, driving winds, or freezing temperatures are likely to occur. On the other hand, weathered joints—a type recommended for weathertight walls—shed water easily but require careful finishing; i.e., they must be worked with a trowel from below.

The overall appearance of a masonry wall depends not only on the joint treatment but also on the color uniformity of the joints. Although mortar shade is influenced to some degree by the moisture condition of the units and by the atmospheric conditions, it depends mainly on the uniformity of the mortar mix and the time of joint tooling. The amount of water used in mixing colored mortar greatly influences the shade and thus requires accurate control. Retempering of colored mortar should be avoided, and any mortar that has become too stiff for use should be discarded. A darker color results if the mortar is tooled when relatively hard rather than reasonably plastic, but some masons consider mortar ready for tooling only when thumbprint-hard. Uniform time of tooling is important for obtaining uniformly colored joints.

Patching and Pointing

In spite of good workmanship, joint patching or pointing may sometimes be necessary. Mortar in a head joint may have fallen out while the units were being placed, or a mortar crack may have formed while the units were being aligned. Furthermore, there may not have been enough mortar in a joint to fill the space left by a broken corner or edge. Sufficient additional mortar should be forced into such spots to completely fill the joints.

Patching or pointing is done preferably while the mortar in the joint is still fresh and plastic. If the back of the face shell can be reached when forcing additional mortar into the joint, the mason provides a back-stop, such as the handle of a hammer. Any depressions and holes made by nails or line pins are filled with fresh mortar before tooling.

When patching or pointing must be done after the mortar has hardened, the joint is chiseled out to a depth of about 1/2 in., thoroughly wetted, and re-pointed with fresh mortar.

Keeping Out Termites

Termites can squeeze through openings as small as 1/32 of an inch and their shelter tubes have been found in the cavities of concrete masonry. Thus, concrete masonry requires the following special precautions beyond those usually made in construction that needs control and protection against these wood-eating insects.

Masonry foundations, piers, and basement walls should be treated with a chemical such as aldrin, dieldrin, or chlordane. The chemical—at least 2 gal. for each 10 lin.ft.—is injected or buried below the surface of the ground along the wall or the base.

In addition, all masonry foundation walls and piers

Fig. 6-48. Raked joints accent a split-block wall.

should be capped with cast-in-place concrete that is at least 4 in. thick and reinforced with two No. 3 bars. This capping should extend the full width of the wall and across the voids in veneered or cavity walls. The top of the cap should be at least 8 in. above grade. Solid masonry units are not acceptable as a substitute for cast-in-place capping.

Bracing Walls

Too frequently a freshly laid masonry wall will be blown over by wind. Such losses can be prevented. Good construction practice, most building codes, and the U.S. Occupational Safety and Health Administration (OSHA) regulations* require that new walls be braced for wind.

Concrete masonry walls usually are not designed to be freestanding. Wind pressure can create four times as much bending stress in a new freestanding wall as in a finished wall of a building. This stress occurs at the bottom of the wall where flashing or lack of bond decreases the wall strength to resist tensile wind forces, and fresh mortar in a wall has little strength. The bracing should be designed to resist wind pressure as required by local regulations for building design.

Bracing should be provided if the height of the wall exceeds that given in Fig. 6-50 for various peak wind velocities. Where bracing is used, the heights shown are the safe heights above the bracing. For example, to withstand wind gusts of 50 mph, the freestanding height (without bracing) of a 10-in.-thick wall weighing 67 psf or less should not exceed 7-1/2 ft. The family of

curves in the figure is based on the assumptions that the mortar has no tensile strength and the walls are freestanding, nonreinforced, ungrouted concrete masonry. In cavity walls the thickness should be assumed to be two-thirds the sum of the thickness of the two wythes.

Quality of Construction

Some time before, during, and after construction questions are asked about the quality of construction. This is investigated by quality control and quality assurance** personnel. It involves several factors:

1. The knowledge and attitude of those in charge determine how much quality is wanted and/or will be achieved.
2. The assignment of responsibility for quality construction depends upon the type of job.
3. The requirements for satisfactory quality in construction are well established but require familiarity.

---

*Sec. 1926.700(a) of the OSHA Construction Safety and Health Regulations adopts, by reference, Article 12.5 of the American National Standard Safety Requirements for Concrete Construction and Masonry Work (ANSI A10.9), which states: "Masonry walls shall be temporarily shored and braced until the designed lateral strength is reached, to prevent collapse due to wind or other forces."

**"Quality control" is the contractor's or manufacturer's effort to achieve a specified result. "Quality assurance" is the owner's effort to measure quality and determine its acceptability.

Fig. 6-49. Newly laid concrete masonry walls are braced against wind because of their low strength when freestanding.

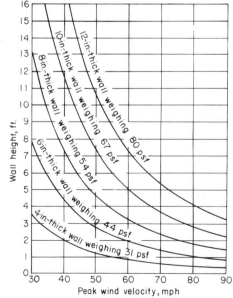

Fig. 6-50. Maximum unsupported heights of nonreinforced, ungrouted concrete masonry walls during construction. Source: Ref. 52.

Fig. 6-51. Construction of a high-rise hotel with engineered load-bearing concrete masonry. Strict quality control is required.

For the owner and his quality assurance representative the key to success—a quality concrete masonry structure—is a reputable masonry contractor. The local masonry association can usually suggest a contractor or answer questions concerning a contractor's reputation. The Better Business Bureau is also helpful along these lines. Another good practice is to check on the contractor's references and previous work.

In the field the roles of the quality control and quality assurance inspectors are much greater than overseeing a few tests and inspecting the completed structure. They must become very familiar with the job specifications and drawings. As this is a lengthy process, it is a great advantage to have the engineer or architect or their authorized representatives on the jobsite.

The quality control inspector must also thoroughly understand the masonry specifications; reflecting the experience of the entire industry, each paragraph and sentence is packed with significance. Reference specifications are equally important, and copies should be at hand on the job for the inspector. Moreover, since a

competent inspector must know or be able to quickly determine many things, it is suggested that inspectors become familiar with this entire handbook. One of its purposes is to explain and set forth recommended practices for quality construction.

Quality and safety of construction can usually be equated, particularly in engineered masonry construction. In this type of work the importance of inspection is so great that allowable stresses should be reduced, according to an ACI Committee 531 report,* if there is no engineering or architectural supervision of the construction. The ACI report also spells out the inspection requirements, stating: "Anything which the inspector feels may be deleterious to the structural strength or architectural appearance of the masonry as set forth in the plans and specifications should be called to the attention of the architect or engineer."

A few special notes follow on physical tests that are often required.

Sampling and Testing Units

Compressive strength, absorption, weight, moisture content, and dimensions are tested on masonry units selected at the place of manufacture from the lots ready for delivery. At least 10 days should be allowed for the completion of the tests. Tests for drying shrinkage, however, must be conducted long before delivery. The allowable moisture content depends on the average annual relative humidity of the locale, and U.S. and Canadian data are given in Fig. 6-52 and Table 6-6, respectively.

Compression tests require testing machines of large capacity. The equipment in most commercial testing laboratories has a maximum capacity of 300,000 lb., while the ultimate strength of a single high-strength masonry unit may greatly exceed this. Therefore, ASTM C140 permits a unit to be "sawed into segments" if a lower-capacity testing machine has to be used. For example, a two-core unit could be cut down to a one-core unit.**

Making a Sample Panel

Occasionally in concrete masonry work a sample panel of the wall must be built by the contractor at the start of the job. It is the responsibility of the quality assurance representative to approve the sample panel; however, the architect or owner himself approves the overall appearance of the panel for important architectural work.

Often 64 in. long and 48 in. high, the sample panel should include the masonry reinforcement as well as

---

*See Ref. 52.
**See Ref. 44.

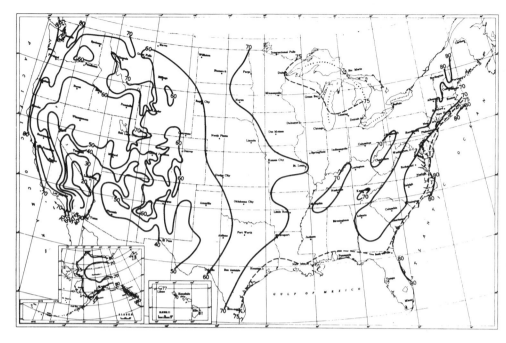

Fig. 6-52. Mean (average) annual relative humidities, percent, for the United States. Prepared by Environmental Science Services Administration, U.S. Department of Commerce, 1968.

Table 6-6. **Mean (Average) Annual Relative Humidities— Canada***

City	Relative humidity, percent
Calgary, Alta.	62
Edmonton, Alta.	68
Prince Rupert, B.C.	85
Vancouver, B.C.	81
Victoria, B.C.	79
Churchill, Man.	80
Winnipeg, Man.	70
Fredericton, N.B.	72
Saint John, N.B.	80
St. Johns Torbay, Newf.	83
Arctic Bay, N.W.T.	75
Fort Smith, N.W.T.	71
Halifax, N.S.	79
Kapuskasing, Ont.	74
London, Ont.	77
Ottawa, Ont.	69
Toronto, Ont.	73
Windsor, Ont.	75
Charlottetown, P.E.I.	81
Montreal, Que.	70
Nitchequon, Que.	79
Quebec, Que.	71
Regina, Sask.	69
Dawson, Y.T.	74

*Based on *Climatic Normals—Vol. 4, Humidity*, Meteorological Branch, Department of Transport, Toronto, Ont., 1968.

show the workmanship, coursing, bonding, thickness, tooling of joints, range of unit texture and color, and mortar color—all as specified or selected. Of course, the finished work must match the sample panel.

Testing Field Mortar

Before construction begins, 2-in. mortar test cubes may be prepared in the laboratory (in accordance with ASTM C270) to check the characteristics of the cementitious materials and sand. These cube tests are made only if the mortar mixture has been specified by strength, not proportions. Because a much greater water content is used during construction, the strengths of any cubes made from mortar on the job cannot be compared to the ASTM strength specifications. However, periodic cube tests made from field mortar can be used to check the uniformity of batching operations. They will be even more useful if they are started with accurately measured batches.

ASTM specification C780 is currently useful for recommending specific masonry mortar field tests and for interpreting the test results. These test results provide a more reliable measure of uniformity of field batching than if periodic cube tests were made according to ASTM C270. Another useful mortar test for field quality control is provided in a 1970 report of ACI Committee 531.*

---

*See Ref. 52.

Testing Wall Strength

Prism strength tests (Fig. 6-53), discussed in Chapter 3 (page 45), may be required not only in advance of design but also as part of quality control in the field, especially on projects involving engineered concrete masonry. During construction, one test is usually required for every 5,000 sq.ft. of wall; each test includes three prisms and not less than three prisms are made per story. By the probabilities laws, some test values may fall below the specified prism strength and this must be evaluated in light of the discussion in Chapter 3.

Separate field tests for compressive strength of grout, discussed in Chapter 2 (page 40), may be another requirement for quality control. Some codes specify a minimum of 1,000 psi at 7 days and 2,000 psi at 28 days, but 28-day test results for grout containing pea gravel may be as high as 3,500 psi.

Destructive Testing of Grouted Walls

Some quality control practices for grouted concrete masonry walls involve the use of destructive testing; i.e., removal of cores drilled through the completed walls and submittal of these cores to visual inspection and physical tests. This practice not only causes damage to the walls but also is expensive, is not directly repeatable, and does not necessarily determine locations of imperfections.

Where grouted walls are to be cored, the cores should be taken at least 6 ft. from a corner and not under an opening in the wall. The governing code or specification should be checked for the size of core required, and at least two cores should be tested for each project. Those codes that require core tests usually specify a minimum compressive strength of 750 psi at 7 days and 1,500 psi at 28 days.

It should be recognized that, while the cores are drilled laterally through the wall, they must be test-loaded along the core axis that is at right angles to the axis of the applied loads on the wall. Therefore, it is unlikely that the test result is a true measure of wall strength. A truer representation of the strength of the masonry wall in the direction of the anticipated applied loads is obtained by testing a sample prism cut out of the wall.

Generally through-the-wall cores are taken to check their compressive strength. However, they may also be taken to check the shear bond strength between the masonry units and grout. The test determines the unit force required to shear the masonry face shells from the grout core for each face. Some building codes specify a minimum shear bond strength of 100 psi at 28 days.

Nondestructive Testing of Grouted Walls

Without damage to the walls, two tests permit

Fig. 6-53. Method of testing the compressive strength of a masonry prism.

considerable wall area to be inspected for major variations or flaws. These tests consist of ultrasonic and radiographic inspection.

The ultrasonic test measures the time that sound waves take to pass through the thickness of a grouted wall. When the time interval indicating good grouting is known, the time measure can be used as a quality control tool. A relatively long time interval indicates the presence of flaws such as voids, cracks, or honeycomb. The moisture content of the wall affects the time interval but does not interfere with identifying variations within the wall.

Ultrasonic testing can also be used to measure the strength of the wall by the amount of sound traveling through it. A large loss of input sound level, easily detected by the almost complete loss of the received signal, indicates large areas of voids, cracks, and honeycomb. Although ultrasonic inspection could cost more than drilling and testing cores, the more complete coverage available may well justify the increased cost.

In the radiographic test, a beam of either X-ray or cobalt 60 isotope radiation is passed through a grouted wall to expose a photographic film on the reverse side. The film shows variations in wall density as gradations of black and white, and thus the sensitivity of the procedure depends to a large extent on the type of wall constructed. Since radiation is involved, a hazard does exist with this method of inspection.

A wide variety of applied finishes are possible with concrete masonry construction. The finish to use in any particular case will be governed by the type of structure in which the walls will be used, the climatic conditions to which they will be exposed, and the architectural effects desired. Some popular finishes are described in this chapter.

Paints

The main purposes of painting concrete masonry walls are to add a fresh appearance and color and to alter the surface texture and pattern. Additional purposes are to reduce the passage of sound through the wall and, as discussed below, to bar the passage of moisture. The paint selected should not only achieve these goals but also attach itself closely to the underlying surface, retain its appearance during a long life, and be economical.

Paints are pigmented coatings that form opaque films. Paint mixtures have minute solid particles of pigment suspended in liquid referred to as the vehicle. The pigment provides hiding power (conceals the surface beneath) and color. The vehicle includes two elements: the volatile solvent or thinner that supplies the desired consistency for application, and the binder that bonds the pigment particles into a cohesive paint film during the drying and hardening process. Some paints dry and harden by evaporation of the solvent; some, by oxidation. Most of the other types involve chemical reactions as well as evaporation of the solvent.

Some paints breathe; that is, they are permeable and allow passage of water vapor. The others are non-breathing or impermeable. Impermeable paints should be applied to the side of a wall where moisture enters; permeable paints, to the surface where moisture exits. If the surface from which moisture is attempting to leave is coated with an impermeable paint, blistering will occur and the paint will peel.

There are several other items that must be considered when selecting paint for a concrete masonry surface. Among them are: (1) whether or not the paint will be damaged by the probable presence of alkalies, and (2) whether the surface texture requires alteration before application of the paint.

Types Commonly Used

Many products have been marketed for use in painting concrete masonry walls and they have had varying degrees of success. The basic constituents and pertinent characteristics of paints now commonly used on concrete masonry walls are described below. Urethanes, polyesters, and epoxies are also used successfully.

Portland Cement Paints

Of the various types of paint used on concrete masonry construction, those with a portland cement base (Fig. 7-1) have the longest service record. U.S. Federal Specification TT-P-21, which gives requirements for composition, provides for two types and two classes of portland cement paint: Type I (containing a minimum of 65% portland cement by weight), for general use; Type II (with at least 80% portland cement by weight), for maximum durability. Under each type, Class A contains no aggregate filler and is for general use, while Class B contains 20 to 40% siliceous sand filler for use on open-textured surfaces.

A concrete masonry surface should be damp at the time of application of a portland cement paint (Fig. 7-2). The setting and curing require the presence of water, a favorable temperature, and sufficient time for hydration. If the paint is modified with latex, moist curing is not necessary because the latex retains

Fig. 7-1. A stippled, heavy cement-based coating on a concrete masonry screen wall.

Fig. 7-2. The concrete masonry surface is uniformly dampened with a water spray just before application of a portland cement paint. A garden pressure sprayer with a fine fog nozzle is recommended.

sufficient moisture in the paint film for hydration.

Although portland cement paints may be made on the job, the best results (uniform color, durability, etc.) are most often secured by using those marketed in prepared form. Portland cement paints form hard, flat, porous films that readily permit passage of water vapor. These paints are not harmed by the presence of alkalies and may be applied to freshly erected concrete masonry surfaces. However, the results will be better if painting is deferred at least three weeks.

Latex Paints

Latex paints are water-thinned; i.e., they are based on aqueous emulsions of various resinous materials such as acrylic resin, polyvinyl acetate, and styrene-butadiene. With an ever increasing use of polymers, blends, and modifications of the base resins, the distinction between these materials has been blurred to the point where any such classification is somewhat meaningless. They all dry very rapidly by evaporation of water, followed by coalescence of the resin particles.

Careful surface preparation is required for latex paints since they do not adhere readily to chalked, dirty, or glossy surfaces. However, these paints are easy to apply and have little odor. Also, they are economical, nonflammable, breathing paints that are not damaged by alkalies. They have excellent color retention and are very durable in normal environments. Those containing acrylics are more expensive than the other latex types, but field experience has shown them to give the best performance.

Oil-Based Paints and Oil-Alkyds

Oil-based paints contain drying oils as the binder and are nonbreathing. They are similar to conventional house paints. Easy to use, these paints are durable under some exposures but not particularly hard or resistant to abrasion, chemicals, or strong solvents. They are also damaged in the presence of alkalies.

Oil-based paints are often modified with alkyd resins to improve resistance to alkalies, reduce drying time, and improve performance in other ways. When the substitution of resins for oil is high, they are referred to as varnish-based paints. The oil-alkyds and even the varnish-based paints may be susceptible to damage from alkalies. Oil-alkyds also are nonbreathing paints.

There are few instances where serious consideration should be given to an oil-based or oil-alkyd paint for use on concrete masonry.

Rubber-Based Paints

Formulated with chlorinated natural rubber or styrene-butadiene resin, these paints form a nonbreathing film with alkali and acid resistance. They are used not only for exterior masonry surfaces but also for interior ones that are wet, humid, or subject to frequent washing (swimming pools, wash and shower rooms, kitchens, and laundries); i.e., where alkali resistance is important and where requirements for resistance to the entrance of moisture are greater than can be supplied by latex paints. Rubber-based paints may be used as primers under less resistant paints.

Surface Preparation

Regardless of the type of paint selected, its success or failure can be dependent upon the adequacy of surface preparation. In most cases success is more assured if the concrete masonry surface can age at least six months before painting; this is due to the dampness usually present in a new masonry wall and to alkalinity of the surface. If the paint is not sensitive to either moisture or alkalies (such as a portland cement paint or latex paint), a long aging period is unnecessary. Earlier use of paint subject to damage from alkalinity is possible if the surface is neutralized by pretreatment with a 3% solution of phosphoric acid followed by a 2% solution of zinc chloride. However, this procedure has become rare because of the successful use of other paints that do not require pretreatment, such as portland cement and latex paints.

For paint to adhere, concrete masonry surfaces must be free of dirt, dust, grease, oil, and efflorescence. Dirt and dust may be removed by air-blowing, brushing, scrubbing, or hosing. If a surface is extremely dirty, wet or dry sandblasting, waterblasting, or steam-cleaning may be used. Grease and oil are removed by applying a 10% solution of caustic soda, trisodium phosphate, or detergents specially formulated for use on concrete. Efflorescence is cleaned off by brushing or light sandblasting. After any of these treatments, the surface should be thoroughly flushed with clean water.

Fill coats, also called fillers or primer-sealers, are sometimes used to fill voids in open or coarse-textured concrete masonry surfaces. Applied by brush before the finish coat(s), the fill coats usually contain white portland cement and fine siliceous sand. If acrylic latex or polyvinyl acetate latex is included in the mixture, moist curing is not required. Fill coats impair sound absorption but improve the sound transmission loss of a concrete masonry wall.

Paint Preparation and Application

Paint must be thoroughly stirred just prior to application. Power stirrers and automatic shakers are becoming more common, but they are not recommended for latex paints because of the possibility of foaming. Hand stirrers should have a broad, flat paddle.

Thinning of paint should only be done in accordance with the manufacturer's directions; excessive paint

Fig. 7-3. Typical brushes used in applying portland cement paint: *(left to right)* ordinary scrub brush, window brush, brush with detachable handle, and fender brush.

Fig. 7-4. Portland cement paint is applied at the joints first.

thinning will result in coatings of low durability. Color tinting should also be done carefully in accordance with the manufacturer's suggestions.

The paints commonly used on concrete masonry are applied by brush, roller, or spray as described below. Roller application is often used for large areas and interior surfaces.

1. *Portland cement paints* are applied to damp surfaces by brush with bristles no more than 2 in. in length (Fig. 7-3). The paint should be *scrubbed* into the surface as shown in Fig. 7-4. An interval of at least 12 hours should be allowed between coats. After completion of the final coat, 48-hour moist curing is necessary if the paint is not modified with latex.

2. *Latex paints* may be applied to dry or damp surfaces by roller or spray, but preferably by long-fiber, tapered nylon brush 4 to 6 in. wide (soaked in water for two hours before use). When the surface is moderately porous or extremely dry weather prevails, it is advisable to dampen the surface. These paints dry throughout as soon as the water of emulsion has evaporated, usually in 30 to 60 minutes, and require no moist curing.

3. *Oil-based and oil-alkyd paints* should not be applied during damp or humid weather or when the temperature is below 50 deg. F. Application is by brush, roller, or spray (usually by brush) to a dry surface. Each coat should be allowed to dry at least 24 hours and preferably 48 hours before application of succeeding coats.

4. *Rubber-based paints* are usually applied by brush to dry or slightly damp surfaces. Two or three coats are necessary to achieve adequate film thickness, and the first coat is usually thinned in accordance with the manufacturer's recommendations. A 48-hour delay is recommended between coats. Recoating should be performed with care. Because of the strong solvents used in rubber-based paints, a second coat tends to lift or attack the original coat.

Caution: Most paints are flammable, and some have solvents that are highly flammable. Therefore, adequate ventilation during painting should be provided according to manufacturers' recommendations.

Clear Coatings

In many areas, architects specify application of a clear coating on concrete masonry structures. This is done to render the surface water-repellent and thus protect the masonry from soiling and surface attack by air-borne pollutants, as well as to facilitate cleaning. Further advantages are to prevent the surface from darkening when wet and to accentuate surface color. In some areas, where weather exposure is not severe and air pollution is low, coatings may not be necessary.

The coating selected should be water-clear and capable of being absorbed into the surface. It should also be long-lasting and not subject to discoloration with time and exposure. Good service has been obtained with coatings based on a methyl methacrylate form of acrylic resin.* Brush or spray application of one or two coats of a relatively low solid-content coating is usually satisfactory.

Stains

Decorative staining of concrete masonry walls can give good results if the proper stain is selected and then applied correctly. However, there may be some drawbacks to staining. For example, color applied after concrete hardens is not as long-lasting as that incorporated into the concrete mix during block manufacture. Also, the color effects or shades of a stain may vary.

Types

Several types of stains may be used on concrete masonry. They include:

1. *Oil stains,* which sometimes require aging of the wall for several months before application of stain, or pretreatment of the surface to inhibit reaction between alkalies and the oil vehicle in the stain. Many stains suitable for wood are suitable for concrete masonry.

2. *Metallic salt stains,* which are slightly acid solutions of salts that result in the deposit of colored metallic oxide or hydroxide in the surface pores. These deposits are not soluble in water.

3. *Organic dyes,* which contain analine dyes or certain of the indicators used in chemical analysis.

Application

For best results, staining should be delayed at least 30 to 45 days after the concrete masonry structure is built. The surface must be dry and clean—free of oil, grease, paint, and wax. Acid etching is usually not necessary or advisable because of the porosity of a concrete masonry surface. Also, two or more stain applications may be required to secure the depth of color desired.

Each coat of stain should thoroughly saturate the surface and be evenly applied by a constant number of passes of a brush, roller, or low-pressure spray. Care should be exercised so that the stain does not streak or overlap into a dried area. From four to five days should elapse between coats, depending upon the masonry surface, ambient conditions, and the stain used. It will often take three or four days for a stained surface to reach its final color.

Portland Cement Plaster Finishes

Portland cement plaster and portland cement stucco are essentially the same finishing material. It can be applied to the surface of any concrete masonry structure. Although stucco is the term often associated with

---

*See Ref. 45.

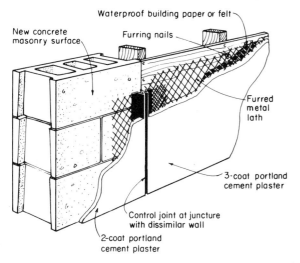

New concrete masonry surface

Waterproof building paper or felt

Furring nails

Furred metal lath

3-coat portland cement plaster

Control joint at juncture with dissimilar wall

2-coat portland cement plaster

Fig. 7-5. Control joint at the juncture of dissimilar walls: *(left)* two-coat plaster applied directly to concrete masonry; *(right)* three-coat plaster on metal lath over wood construction.

exterior use of the material, the term plaster does not explicitly denote either interior or exterior use. Plaster is a combination of portland cement, masonry cement, or plastic cement with sand, water, and perhaps a plasticizing agent such as lime. A color pigment may be used in the finish coat.

Portland cement plaster has many of the desirable properties of concrete and, when properly applied, forms a durable, hard, strong, and decorative finish. There is an unlimited variety of textures, colors, and patterns possible. The rougher textures help conceal slight color variations, lap joints, uneven dirt accumulation, and streaking.

Primarily portland cement plaster finishes are used for exterior walls, but they are also particularly well suited for interior high-moisture locations such as kitchens, laundries, saunas, bathrooms, and various industrial facilities. The use of exterior plaster involves considerations of water penetration, corrosion of reinforcement and accessories, and stresses due to wider variations of temperature and humidity than normally present in interiors. Therefore, exterior applications require practices and precautions that are somewhat different than those for interior applications.

A full discussion of portland cement plastering appears in the *Plasterer's Manual.** Also, extensive details of materials, mixtures, and construction are covered by American National Standards Institute specifications (ANSI A42.2, Standard Specifications for Portland Cement and Portland Cement-Lime Plastering, Exterior [Stucco] and Interior, and A42.3, Standard Specifications for Lathing and Furring for Portland Cement and Portland Cement-Lime Plastering, Exterior [Stucco] and Interior). A brief review of portland cement plaster follows.

Jointing for Crack Control

The proper design and selection of materials for a concrete masonry wall can substantially minimize or eliminate undesirable cracking of its portland cement plaster finish. Cracks can develop in this finish through many causes or combinations of causes; e.g., shrinkage stresses; building movements; foundation settlements; restraints from lighting and plumbing fixtures, intersecting walls or ceilings, pilasters, and corners; weak sections due to cross-section changes such as at openings; and construction joints.

It is difficult to anticipate or prevent cracks from all these possible causes, but they can be largely controlled by means of metal control joints (Fig. 7-5). Whether plaster is applied directly to a concrete masonry base or on furred lath (metal reinforcement),** control joints should be installed directly over any previous joints in the base.

Concrete masonry walls and ceilings that use metal lath for the plaster base should be divided into rectangular panels with a control joint at least every 20 ft. The metal reinforcement in the plaster must not extend across these control joints, and the metal used for the control joints on exterior surfaces should be weathertight and corrosion-resistant.

Mixes

A good plastering mix will be recognized by its workability, ease of troweling, adhesion to bases, and ability to attach itself to surfaces without sagging. Batch-to-batch uniformity will help assure uniform suction for subsequent coats and color uniformity.

Data for plaster mixes are given in Tables 7-1, 7-2, and 7-3. Note that lime should not be added when masonry cement or plastic cement is used. These cements already contain plasticizers and only sand and water need be added, thus simplifying jobsite proportioning and mixing.

Uniform measuring and batching methods are important. All ingredients should be thoroughly mixed (preferably in a power mixer) with the amount of water needed to produce a plaster of workable consistency. Mixing time should be a minimum of 2 minutes after all materials are in the mixer, or until the mix is uniform in color. The size of a batch should be that which can be used immediately or in no more than 2-1/2 hours. Remixing, to restore plasticity with the

---

*See Ref. 18.
**Furring is a term applied to spacer elements used to maintain a gap between lath (or a finish such as wallboard) and the masonry. Lath is a material whose primary function is to serve as a plaster base; the most common types are expanded metal lath, expanded stucco mesh, hexagonal wire mesh (stucco netting), stucco mesh, and welded-wire fabric.

Table 7-1. Permissible Mixes for Portland Cement Plaster Base Coats*

Type of plaster base	Plaster mix symbols	
	First coat (Scratch coat)	Second coat (Brown coat)
Concrete masonry**	L M P	L M P
Metal rein-forcement†	C L CM M CP P	C, L, M or CM L CM or M M CP or P P

*Adapted from ANSI A42.2, Table 2.
**High-absorption bases such as concrete masonry should be moistened prior to scratch coat application.
†Metal reinforcement with paper backing may require dampening of paper prior to application of plaster.

Table 7-2. Proportions for Portland Cement Plaster Base Coats*

Plaster mix symbols	Cementitious materials, parts by volume				Sand**	
	Portland cement	Lime**	Masonry cement	Plastic cement	First coat	Second coat†
C	1	0 to 3/4	—	—	Not less than 2-1/2 and not more than 4 times the sum of the volumes of cementitious materials used	Not less than 3 and not more than 5 times the sum of the volumes of cementitious materials used
CM	1	—	1 to 2	—		
L	1	3/4 to 1-1/2	—	—		
M	—	—	1	—		
CP	1	—	—	1		
P	—	—	—	1		

*Adapted from ANSI A42.2, Table 1.
**Variations in lime and sand contents are given due to variations in local sands, and the fact that higher lime content will permit use of higher sand content. The workability of the plaster mix will govern the amounts of lime and sand.
†Within the limits shown, the same or greater proportions of sand should be used in the second coat as in the first coat.

Table 7-3. Proportions for Job-Mixed Portland Cement Plaster Finish Coat*

Plaster mix symbols	Cementitious materials, parts by volume				Sand
	Portland cement	Lime	Masonry cement	Plastic cement	
F**	1	3/4 to 1-1/2	—	—	Not more than 3 times the sum of the volumes of cementitious materials used
FL	1	1-1/2 to 2	—	—	
FP	—	—	—	1	
FM	—	—	1	—	

*Adapted from ANSI A42.2, Table 3.
**Specify for surfaces subjected to abrasive treatment.

addition of water, is permissible within the same time limits.

Color pigments are often used in the finish coat, which is usually a factory-prepared stucco* finish mix. It should be noted that factory-prepared finish mixes assure greater uniformity of color than job-prepared mixes, and the manufacturer's recommendations should be closely followed. If the finish coat is job-mixed, truer color and a more pleasing appearance will be obtained when a white portland cement and a fine-graded, light-colored sand are used.

Surface Preparation

Concrete masonry provides an excellent base for plaster because of its rigidity and excellent bonding characteristics. Bond occurs both mechanically and chemically. Mechanically, bond results from keying; i.e., interlocking of plaster with the open texture in the concrete masonry surface. Suction by the masonry also improves mechanical bond; i.e., plaster paste is drawn into minute pores of the surface. Chemically, the similar materials adhere well to each other.

A new concrete masonry surface can be used as a plaster base with minimum consideration. For best results, the concrete masonry units should have an open texture and be laid with struck joints. The surface should be free of oil, dirt, or other materials that reduce bond; then prior to application of plaster, it is uniformly dampened, but not saturated, with clean water (Fig. 7-7).

An old concrete masonry surface should be as-

*In some areas the word stucco refers only to the finish coat while in most other areas it refers to the entire thickness.

sessed to establish its bonding characteristics. A surface having the desired texture and cleanliness will perform as well as a new masonry surface. If the masonry has been painted, sandblasting must be used to remove the paint and improve the bonding characteristics. Otherwise, the open surface texture required must be obtained by anchoring metal lath to the surface over waterproof building paper or felt (Fig. 7-8).

The suitability of a concrete masonry surface as a plaster base can be tested by spraying it with clean water to see how quickly moisture is absorbed through suction. If water is readily absorbed, good suction is likely; if water droplets form and run down the surface, its suction is probably inadequate. For low-suction surfaces, bond must be increased by applying

Fig. 7-7. A water spray is used to prepare the concrete block surface for its first coat of plaster.

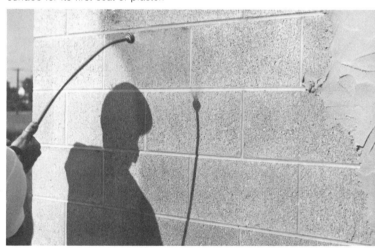

Fig. 7-8. Hand application of plaster on metal lath over waterproof building paper or felt. Dimples in the expanded metal lath provide proper furring for embedment of the metal.

Fig. 7-6. The natural rough texture of concrete block provides a good base for portland cement plaster. A suitable corner bead is provided by metal corner reinforcement with a series of coiled wires.

a bonding agent or a dash-bond coat (containing 1 part portland cement, 1 to 2 parts sand, and sufficient water for a thick paint-like consistency). In lieu of a bonding agent or dash coat, bond must be provided by metal lath over waterproof building paper or felt.

Application

Plaster may be applied, by hand or machine, in two or three coats in accordance with the required thicknesses given in Table 7-4. Horizontal (overhead) application seldom exceeds two coats, and two coats are often used when plaster is applied directly to concrete masonry, as shown in Fig. 7-9. Three coats are applied when metal lath is used as a plaster base. A sprayed color coat—the finish coat—requires two applications, the first to ensure complete coverage and the second to obtain the desired texture.

Hand application (Fig. 7-8) involves traditional plastering tools and practices proven successful over many years. However, machine application (Fig. 7-10) is now widely used because it offers many advantages: pumping capability, speedier application, elimination of lap and joint marks, possibility of deeper and darker colors, more uniform texture, and a wider variety of possible textures.

Machine application requires different procedures than those used for hand application. Basically, the mix is sprayed from the machine nozzle against the prepared base or previous coat. Nevertheless, if machines are to be used to best advantage, manufacturers' instructions should be carefully followed.

Sloppy (wet) mixes should be avoided, even though they may go through the hose more easily. Overwatering may cause the plaster materials to separate and result in such problems as glazing, shrinkage cracking, dropouts, low strength, efflorescence, and crazing.

Proper consistency for machine-applied plaster is best determined by observing plaster during application. If the plaster is too fluid to build up the proper thickness, the water content is too high; if plaster will not pump or is exceedingly difficult to strike off, the water content is too low. Sometimes specifications require a slump test for plaster taken from the nozzle of the plastering hose. Maximum allowable slump should be 2-1/2 in. when tested in a 2×4×6-in.-high slump cone.

Curing

Portland cement plaster requires moist curing after application in order to produce a strong, durable finish. Curing can be successfully accomplished by several methods, and the type and size of the structure and the climatic and job conditions will dictate the most suitable. For example, plaster can be moist-cured by using a suitable covering, such as polyethylene

film, to provide a vapor barrier that will retain the moisture in the plaster. Moist curing can also be accomplished by using a fine spray of water. Tarpaulins or plywood barriers that deflect sunlight and wind will help to reduce evaporation rates.

An adequate temperature level is important to satisfactory curing. As the temperature drops, hydration slows and practically stops when the temperature approaches the freezing point. Therefore, portland cement plaster should not be applied to frozen surfaces, and frozen materials should not be used in the mix. Furthermore, in cold weather it may be necessary to heat the mixing water and the work area. ANSI

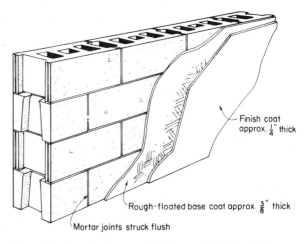

Fig. 7-9. Portland cement plaster applied directly to concrete masonry.

Finish coat approx. $\frac{1}{4}$" thick

Rough-floated base coat approx. $\frac{3}{8}$" thick

Mortar joints struck flush

Fig. 7-10. Machine application of plaster.

Table 7-4. Thicknesses of Portland Cement Plaster*

Type of work		Type of plaster base		Coat thickness, in.							
				Coats on vertical surface				Coats on horizontal surface			
				1st	2nd	3rd**	Total	1st	2nd	3rd**	Total
Interior	3-coat work†	Metal reinforcement		1/4	1/4	1/8	5/8	1/4	1/4	1/8	5/8
		Solid base	Unit masonry	1/4	1/4	1/8	5/8	Use 2-coat work			
			Metal reinforcement over solid base	1/2	1/4	1/8	7/8	1/2	1/4	1/8	7/8
	2-coat work††	Solid base	Unit masonry	3/8	1/4	—	5/8	—	—	—	3/8 max.‡
Exterior	3-coat work†	Metal reinforcement		3/8	3/8	1/8	7/8	1/4	1/4	1/8	5/8
		Solid base	Unit masonry	1/4	1/4	1/8	5/8	Use 2-coat work			
			Metal reinforcement over solid base	1/2	1/4	1/8	7/8	1/2	1/4	1/8	7/8
	2-coat work††	Solid base	Unit masonry	3/8	1/4	—	5/8	—	—	—	3/8 max.‡

*Adapted from ANSI A42.2, Table 4. Where a fire rating is required, plaster thickness should conform to the applicable building code or to an approved test assembly.

**The finish coat thickness may vary, provided that the total plaster thickness complies with this table and is sufficient to achieve the texture specified. For exposed-aggregate finishes, the second (brown) coat may become the "bedding" coat; it should be of sufficient thickness to receive and hold the aggregate specified.

†Where three-coat work is required, dash or brush coats of plaster are not acceptable as one of the three coats.

††For two-coat work, only the first and finish coats for vertical surfaces and the total plaster thickness for horizontal surfaces are indicated. The use of two coats is common practice when plaster is applied directly to vertical concrete masonry, and horizontal application seldom exceeds two coats.

‡On horizontal solid-base surfaces such as ceilings or soffits requiring more than 3/8-in. plaster thickness to obtain a level plane, metal reinforcement should be attached to the concrete masonry and the thickness specified for three-coat work on metal reinforcement over solid base applies. Where 3/8-in. or less plaster thickness is required to level and decorate and there are no other requirements, a liquid bonding agent or dash-bond coat may be used.

Specification A42.2 requires a heated enclosure to maintain the air temperature above 40 deg. F. for 48 hours before plastering, while plastering, and for the duration of curing (at least 48 hours).

Furred Finishes

In addition to plaster, other finishes such as fiberboard, gypsum wallboard, and wood paneling are sometimes applied directly to concrete masonry wall surfaces (Chapter 3) and sometimes upon furring. The furring, which may be wood or metal, ensures a definite air space 3/4 to several inches wide between the masonry and the finish (Fig. 7-11). Furring may be necessary to:

1. Provide suitably plumb, true, and properly spaced supporting construction for a wall finish.
2. Eliminate capillary moisture transfer in exterior or below-grade concrete masonry walls, thus minimizing the likelihood of condensation on interior wall surfaces.
3. Improve thermal insulation.
4. Improve sound insulation.

The furring strips are fastened to the concrete

Fig. 7-11. Nailing wallboard to wood furring strip.

Fig. 7-12. Wood furring strips are nailed to the mortar joints between the block.

masonry with cut nails, helically threaded concrete nails, or powder-actuated fasteners (Fig. 7-12). Adhesives can be used to attach wallboard directly to wood furring, although a few nails may be required until the adhesive has set.

To comply with Canadian fire rating requirements, lath and wallboard must be securely fastened to the furring as described in Chapter 3 (page 69). It should be noted that wood furring is combustible even if enclosed by masonry and a fire-resistant finish. Building codes sometimes prohibit such combustible construction.

Gypsum wallboard finishes consist of one or more plies of factory-fabricated gypsum board having noncombustible gypsum cores with surfaces and edges of paper or other materials. Gypsum plaster may be applied directly to concrete masonry or to lath that may or may not be furred. Gypsum products are not recommended where significant exposure to moisture is expected.

Typical materials used to increase thermal resistance of furred finishes include: (1) flexible fiber insulation (batts and blankets), (2) loose-fill insulation, (3) rigid plastic insulation, and (4) reflective-foil insulation attached to the back surface of wallboard. Mineral fiber blankets may be installed behind the lath or wallboard not only to improve thermal insulation but also to decrease sound transmission. Resilient clips for lath attachment may also be used to decrease sound transmission.

The most common application of concrete masonry is for built-in-place walls for buildings of all kinds. However, there are a number of other common applications, as described in this chapter.

Fireplaces and Chimneys

Fireplaces and chimneys are important elements in the design and construction of a home. The fireplace can be a central feature for family social life, and the chimney often is a dominant and interesting architectural feature on the home exterior as well as a focal point of interior design. Accordingly, it is desirable that fireplaces and chimneys be esthetically pleasing as well as functional.

Various requirements for fireplaces and chimneys normally are set forth in local building codes, but they are usually for a single residential fireplace with the chimney tied to the roof or ceiling rafters. In the event the chimney is multistory, extra wide, or extra high or there are multiple fireplaces and flues within the

Fig. 8-1. Attractive single-face fireplace of split-face concrete masonry units.

chimney, special design considerations are necessary. A fireplace can be located on one floor directly above one on the floor below, but each fireplace must have a separate flue. Fig. 8-2 shows a way to combine multi-level fireplaces into one chimney. Each flue takes off properly from the center of the smoke chamber.

The design and construction of an efficient, functional fireplace requires adherence to basic rules concerning fireplace location and the dimensions and placement of various component parts, keeping in mind the basic functions of a fireplace. These functions are: (1) to assure proper fuel combustion, (2) to deliver smoke and other products of combustion up the chimney, (3) to radiate the maximum amount of heat, and (4) to provide an attractive architectural feature. It must also afford simplicity and firesafety in construction.

Combustion and smoke delivery depend mainly upon the shape and dimensions of the combustion chamber, the proper location of the fireplace throat and the smoke shelf, and the ratio of the flue area to the area of the fireplace opening. The third objective, heat radiation, depends on the dimensions of the combustion chamber. Firesafety depends not only on the design of the fireplace and chimney but also on the ability of the masonry units to withstand high temperatures without warping, cracking, or deteriorating.

Types of Fireplaces

There are several types of fireplaces being used today, and the basic principles involved in their design and construction are the same. These types and some standard sizes found to work satisfactorily under most conditions are given in Table 8-1.

The *single-face* fireplace (pictured in Fig. 8-1) is the oldest and most common variety, and most standard design information is based on this type. The *multi-face* fireplace, used properly, is highly effective and attractive, but it may present certain problems as to draft and opening size that usually must be solved on an individual basis. The *two-face (opposite)* fireplace is not recommended by some governmental agencies. If used, this type of fireplace requires a fire screen of fire-resistant, tempered pyrex glass to be placed on one side to prevent fire from blowing out into the room.

Barbecues or outdoor fireplaces can discharge into a chimney attached to the house with a separate flue or, if desired, can be located separately from the house with its own chimney. Inexpensive, serviceable barbecues can be built of concrete masonry with minimum labor and time. The site selected should be sheltered from the wind, conveniently located between play and work areas, and afford adequate entertaining space. Details of a simple barbecue are shown in Fig. 8-4 and those of a more elaborate one in Fig. 8-5.

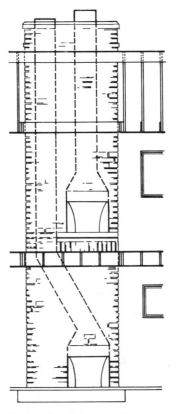

Fig. 8-2. Chimney with separate flues for fireplaces on two floors.

Fig. 8-3. A concrete masonry barbecue enhances a patio or backyard for friendly entertainment.

Table 8-1. Fireplace Types and Standard Sizes*

Type	Width (w), in.	Height (h), in.	Depth (d), in.	Area of fireplace opening, sq.in.	Nominal flue sizes (based on 1/10 area of fireplace opening),** in.
Single-face	36	26	20	936	12×16
	40	28	22	1,120	12×16
	48	32	25	1,536	16×16
	60	32	25	1,920	16×20
Two-face (adjacent)	39	27	23	1,223	12×16
	46	27	23	1,388	16×16
	52	30	27	1,884	16×20
	64	30	27	2,085	16×20
Two-face (opposite)	32	21	30	1,344	16×16
	35	21	30	1,470	16×16
	42	21	30	1,764	16×20
	48	21	34	2,016	16×20
Three-face	39	21	30	1,638	16×16
	46	21	30	1,932	16×20
	52	21	34	2,184	20×20

*Adapted from Ref. 15. The types are illustrated below.
**A requirement of the U.S. Federal Housing Administration if the chimney is 15 ft. high or over; 1/8 ratio is used if chimney height is less than 15 ft. See Table 8-2 for nominal and actual flue sizes and inside areas of flue linings.

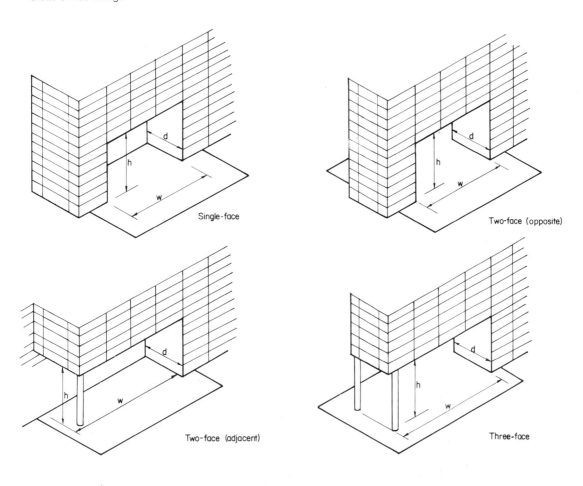

Single-face

Two-face (opposite)

Two-face (adjacent)

Three-face

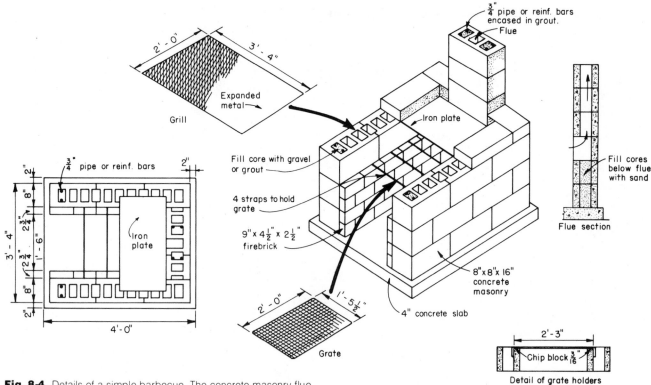

Fig. 8-4. Details of a simple barbecue. The concrete masonry flue must be kept at least 24 in. clear of any combustible construction.

Fireplace Elements

Only the major elements of a fireplace are discussed below, but the details of a typical unreinforced concrete masonry fireplace are shown in Fig. 8-6.

Hearth

The floor of the fireplace is called the hearth. The inner part of the hearth is lined with firebrick and the outer hearth consists of noncombustible material such as firebrick, concrete brick, concrete block, or just concrete. The outer hearth is supported on concrete that may be part of a ground floor or a cantilevered section of the slab supporting the inner hearth.

Lintel

The fireplace lintel is the horizontal member that supports the front face or breast of the fireplace above the opening. It may be made of reinforced masonry or a steel angle, the same as other lintels discussed in Chapter 4. However, it is possible to eliminate the lintel by use of a masonry arch.

Firebox

The combustion chamber where the fire occurs is called the firebox. Its sidewalls are slanted slightly to radiate heat into the room, and its rear wall is curved or inclined to provide an upward draft to the throat (described below).

Unless the firebox is of the metal preformed type (at least 1/4 in. thick), it should be lined with firebrick that is at least 2 in. thick and laid with thin joints of fireclay (refractory) mortar. Its back and sidewalls, including lining, should be at least 8 in. thick to support the weight of the chimney above.

In the generally accepted method of construction on a concrete slab, the fireplace is laid out and its back constructed to a scaffold height of approximately 5 ft. before the firebox is constructed and backfilled with tempered mortar and brick scraps. Masons should not backfill solidly behind the firebox wall but slush the mortar loosely to allow for some expansion of the firebox.

Throat

The throat of a fireplace is the slot-like opening directly above the firebox through which flames, smoke, and other combustion products pass into the smoke chamber. Because of its effect on the draft, the throat must be carefully designed to be not less than 6 in. and preferably 8 in. above the highest point of the fireplace opening, as shown in Fig. 8-6.

The sloping or inclined back of the firebox should extend to the same height as the throat and form the support for the hinge of a metal damper placed in the

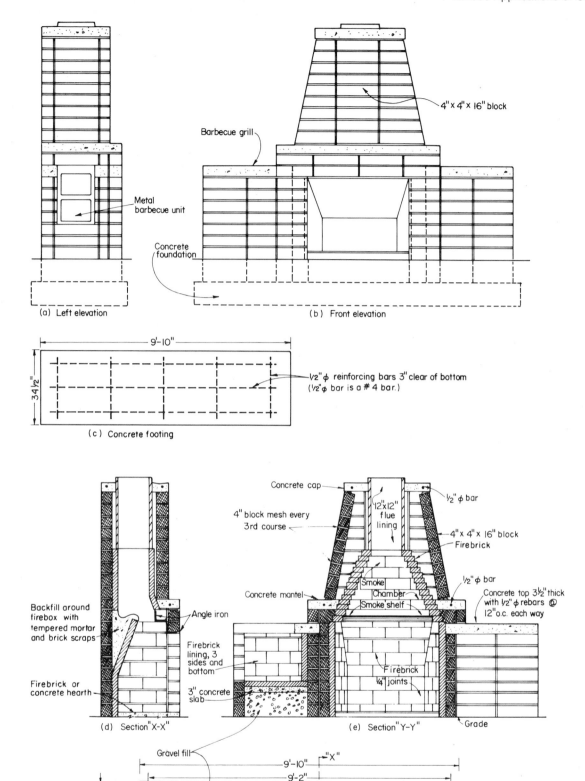

(a) Left elevation

Metal barbecue unit

Barbecue grill

4" x 4" x 16" block

Concrete foundation

(b) Front elevation

9'-10"

34½"

½"φ reinforcing bars 3" clear of bottom
(½"φ bar is a #4 bar.)

(c) Concrete footing

Backfill around firebox with tempered mortar and brick scraps

Firebrick or concrete hearth

Angle iron

Firebrick lining, 3 sides and bottom

3" concrete slab

(d) Section "X-X"

Concrete cap

4" block mesh every 3rd course

12"x12" flue lining

½"φ bar

4" x 4" x 16" block
Firebrick

Concrete mantel

Smoke Chamber
Smoke shelf

½"φ bar

Concrete top 3½" thick with ½"φ rebars @ 12"o.c. each way

Firebrick ¼" joints

Grade

(e) Section "Y-Y"

Gravel fill

"X"

9'-10"

9'-2"

4"

34½" 24½"

"Y"

Concrete

Concrete

"Y"

6"

Wood bin

2'-4"   9"   36"   9"   2'-4"

"X"

(f) First course layout

Fig. 8-5. Details of an elaborate barbecue.

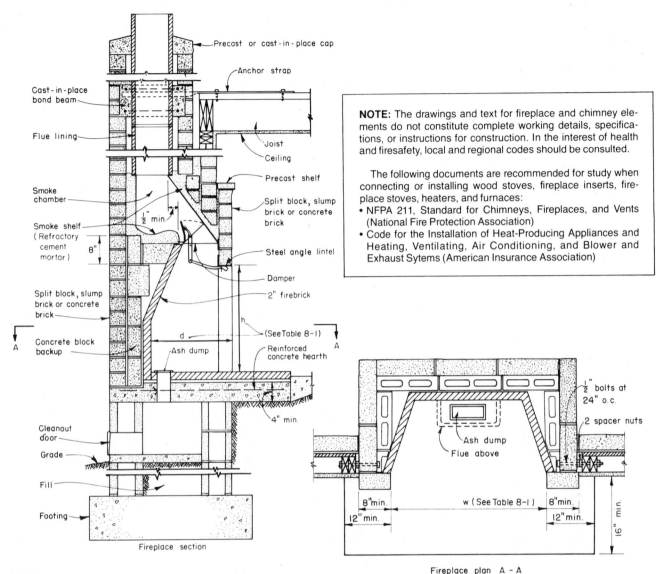

Precast or cast-in-place cap

Anchor strap

Cast-in-place bond beam

Flue lining

Smoke chamber

Smoke shelf (Refractory cement mortar)

Split block, slump brick or concrete brick

Concrete block backup

Cleanout door

Grade

Fill

Footing

Joist

Ceiling

Precast shelf

Split block, slump brick or concrete brick

Steel angle lintel

Damper

2" firebrick

(See Table 8-1)

Reinforced concrete hearth

Ash dump

$\frac{1}{2}$" min.

8"

d

h

4" min.

A

A

Fireplace section

$\frac{1}{2}$" bolts at 24" o.c.

2 spacer nuts

Ash dump

Flue above

8"min.

12" min.

w (See Table 8-1)

8"min.

12" min.

16" min.

Fireplace plan A - A

NOTE: The drawings and text for fireplace and chimney elements do not constitute complete working details, specifications, or instructions for construction. In the interest of health and firesafety, local and regional codes should be consulted.

The following documents are recommended for study when connecting or installing wood stoves, fireplace inserts, fireplace stoves, heaters, and furnaces:
• NFPA 211, Standard for Chimneys, Fireplaces, and Vents (National Fire Protection Association)
• Code for the Installation of Heat-Producing Appliances and Heating, Ventilating, Air Conditioning, and Blower and Exhaust Sytems (American Insurance Association)

Fig. 8-6. Unreinforced concrete masonry fireplace and chimney. Some building codes require that the concrete masonry units be solid units (see definition on page 8).

throat. The damper extends the full width of the fireplace opening and preferably opens both upward and backward.

Smoke Chamber

The smoke chamber acts as a funnel to compress the smoke and gases from the fire so that they will enter the chimney flue above. The shape of this chamber should be symmetrical with respect to the centerline of the firebox in order to assure even burning across the width of the firebox. The back of the smoke chamber is usually vertical and its other walls are inclined upward to meet the bottom of the chimney flue lining. If the wall thickness is less than 8 in. of solid masonry, the smoke chamber should be parged with 3/4 in. of

fireclay mortar. Metal lining plates are available to give the chamber its proper form, provide smooth surfaces, and simplify masonry construction.

Chimney Elements

A fireplace chimney serves a dual purpose: to create a draft and to dispose of the products of combustion. Careful consideration must be given to chimney design and erection in order to assure efficient operation and freedom from fire hazards.

To prevent upward draft from being neutralized by downward air currents, the chimney should be built at least 3 ft. above a flat roof, 2 ft. above the ridge of a pitched roof, or 2 ft. above any part of the roof within a 10-ft. radius of the chimney. If the chimney does not draw well, increasing its height will improve the draft. A typical unreinforced concrete masonry chimney is shown with the fireplace in Fig. 8-6.

Foundation

Usually made of concrete, the foundation for a chimney is designed to support the weight of the chimney and any additional load, such as the fireplace and floors. Because of the large mass and weight imposed, it is important the unit bearing pressure beneath the chimney foundation be approximately equal to that beneath the house foundation; this will minimize the possibility of differential settlement. The chimney foundation is generally unreinforced, with only the chimney reinforcement (where required by local building codes) extending from it.

The footing thickness should be at least 8 in. and not less than 1-1/2 times the footing projection, unless reinforced. The bottom of the footing should be at least 18 in. below grade and extend below the frost line.

Chimney Flue

A fireplace chimney flue must have the correct area and shape to produce a proper draft.* Relatively high velocities of smoke through the throat and flue are desirable. Velocity is affected by the flue area, the firebox opening area, and the chimney height.

Generally the required cross-sectional area of the flue should be approximately 1/10 of the area of the fireplace opening. However, since some codes may specify 1/8 or 1/12 under varying conditions, the local building department should be consulted. Typical sizes of fireplace flues and flue linings are given in Tables 8-1 and 8-2.

A fireplace chimney can contain more than one flue, but each flue must be built as a separate unit entirely free from the other flues or openings. Flue walls should have all joints completely filled with mortar. All chimney flues should be lined. Clay flue liners are the common requirement and are covered by ASTM C315, Standard Specification for Clay Flue Linings. Concrete flue liners made with perlite aggregate and portland cement have been approved in Research Committee Recommendation Report No. 2602 of the International Conference of Building Officials (ICBO).

Flue linings should start at the top of the smoke chamber and extend continuously to 4 to 8 in. above the chimney cap. The chimney walls are constructed around the flue lining segments, which are embedded one upon the other in a refractory mortar such as fireclay, and left smooth on the inside of the lining. Liners should be separated from the chimney wall and the space between the liner and masonry is not filled; only enough mortar should be used to make a good joint and

Table 8-2. Clay Flue Lining Sizes*

Nominal size, in.	Manufactured size (modular), in.**	Inside area, sq.in.
4×8	3-1/2×7-1/2	15
4×12	3-1/2×11-1/2	20
4×16	3-1/2×15-1/2	27
8×8	7-1/2×7-1/2	35
8×12	7-1/2×11-1/2	57
8×16	7-1/2×15-1/2	74
12×12	11-1/2×11-1/2	87
12×16	11-1/2×15-1/2	120
16×16	15-1/2×15-1/2	162
16×20	15-1/2×19-1/2	208
20×20	19-1/2×19-1/2	262
20×24	19-1/2×23-1/2	320
24×24	23-1/2×23-1/2	385

*Source: Clay Flue Lining Institute
**Actual dimensions may vary somewhat, but the flue lining must fit into a rectangle corresponding to the nominal flue size.

hold the liners in position.

Modular-size units (Fig. 8-7) can be combined with modular-size flue lining for modular-size concrete masonry construction. Minimum wall thickness measured from the outside of the flue lining should be 4 in. nominal. The exposed joints inside the flue are struck smooth and the interior surface is not plastered.

Smoke pipe connections should enter the side of the flue at a thimble or flue ring that is built of fireclay or firebrick set with fireclay mortar. The metal smoke pipe should not extend beyond the inside face of the flue, and the top of the smoke pipe should be not less than 18 in. below the ceiling. No wood or combustible materials should be placed within 6 in. of the thimble.

When a chimney contains more than two flues, they should be separated into groups of one or two flues by 4-in.-thick masonry bonded into the chimney wall, or the joints of the adjacent flue linings should be staggered at least 7 in. The tops of the flues should have a height difference of 2 to 12 in. to prevent smoke from pouring from one flue into another. A fireplace on an

*This discussion of flues deals only with residential fireplace chimneys. Commercial and industrial chimneys have more stringent requirements.

upper level should have the top of its flue higher than the flue of a fireplace on a lower level.

For reasons of appearance, chimneys are often built to the same widths as attached fireplaces, and these wide chimneys sometimes contain only a single flue. It can be located anywhere within the chimney. Consideration should be given to reinforcing a wide chimney wall against lateral forces (see discussion "Reinforcement and Chimney Anchorage").

Practically any size or shape of single- or multiple-flue chimney can be constructed with only three different sizes of solid concrete block units (designated Nos. 1, 2, and 3 in Fig. 8-7).

Chimneys should be built as nearly vertical as possible, but a slope is allowed if the full area of the flue is maintained throughout its length. When a slope from the vertical is required in the flue for design reasons, it should not exceed 7 in. to the foot or 30 deg. Where offsets or bends are necessary, they should be formed by mitering both ends of abutting flue liner sections equally; this prevents reduction of the flue area.

Chimney Cap and Hood

The top of the chimney wall should be protected by a concrete cap conforming with the architectural design of the building. The cap should slope not only to

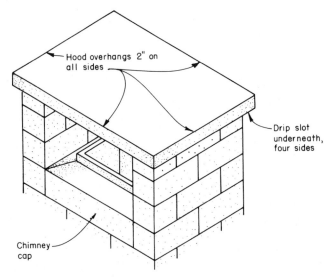

Fig. 8-8. A simple chimney hood keeps rain and snow out, prevents downdrafts, and improves appearance.

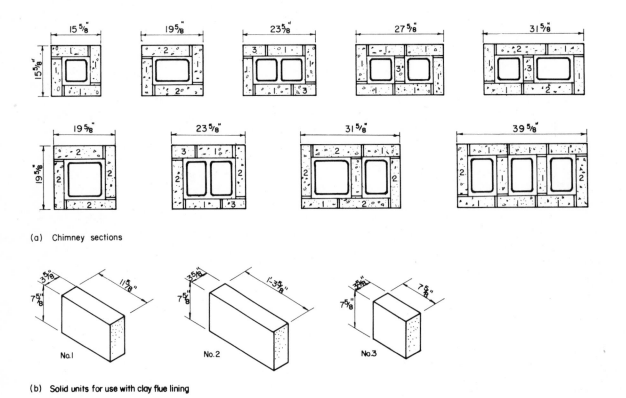

(a) Chimney sections

(b) Solid units for use with clay flue lining

Fig. 8-7. Residential concrete masonry chimney sections. See Table 8-2 for dimensions of clay flue linings.

prevent water from running down next to the flue lining and into the fireplace but also to prevent standing water from creating frost or moisture problems. In addition, since chimney flues should project at least 2 in. above the cap, a sloping cap improves draft from the flue as well as the smoke exhaust characteristics of the chimney. If the cap projects beyond the chimney wall a few inches, a drip slot in its lower edge should be included to help keep the wall dry and clean.

A chimney hood gives a finished touch to the silhouette of the building. It protects the chimney and fireplace from rain and snow and, when the building is located below adjoining buildings, trees, and other obstacles, prevents downdrafts. It must have at least two sides open, with the open areas larger than the flue area. A simple concrete hood is shown in Fig. 8-8.

Concrete chimney hoods should be reinforced with steel bars or welded-wire fabric. If a hood projects from a chimney wall, a drip slot under the edge is included. Also, the openings are sometimes enclosed with heavy screening to keep out small animals and birds, but insurance company regulations on screening should be checked.

Reinforcement and Chimney Anchorage

Depending on local building codes, fireplaces and chimneys have to be reinforced in areas subject to earthquakes or high wind loads. A typical reinforced concrete masonry fireplace and chimney are shown in Fig. 8-9.

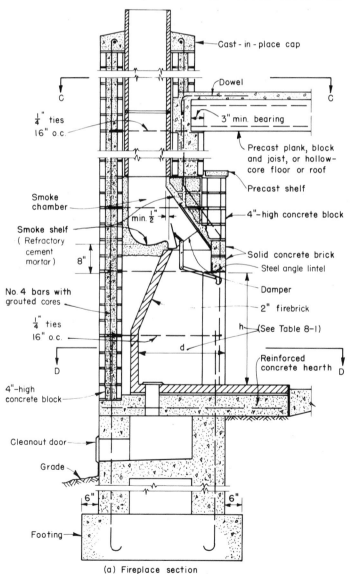

(a) Fireplace section

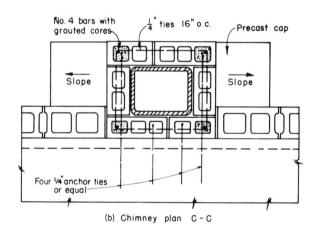

(b) Chimney plan C–C

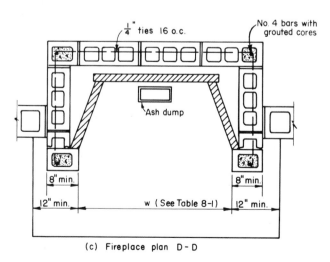

(c) Fireplace plan D–D

Fig. 8-9. Reinforced concrete masonry fireplace and chimney.

The reinforcement, consisting of at least four 1/2-in.-diameter deformed vertical bars, should extend the full height of the chimney and be tied into the footing and chimney cap. Also, the bars should be tied horizontally with 1/4-in.-diameter ties at not more than 18-in. intervals. If the width of a reinforced chimney exceeds 40 in., two additional 1/2-in. vertical bars should be provided for each additional flue incorporated into the chimney, or for each additional 40-in. width or fraction thereof.

All chimneys not located entirely within the exterior walls of a residence should be anchored to the building at each floor or ceiling line 6 ft. or more above grade and at the roof line. The anchors should consist of 1/4-in. steel straps wrapped around vertical reinforcement or chimney flues, as shown in Fig. 8-6. Each end of the strap is attached to the structural framework of the building with six 16d nails, two 1/2-in.-diameter bolts, or two 3/8-in.-diameter by 3-in.-long lag screws. Reinforced chimneys must have equivalent anchorage, as shown in Fig. 8-9.

When a chimney extends considerably above the roof level, an intermediate lateral support or tie is often placed between the roof line and the chimney.

Screen Walls

Concrete masonry screen walls (walls constructed with over 25% exposed open areas) are functional, decorative building elements. They combine privacy with a view, interior light with shade and solar heat reduction, and airy comfort with wind control. Curtain walls, sun screens, decorative veneers, room dividers, and fences are just a few of the many applications of the concrete masonry screen wall.

Materials

With conventional concrete block or the specially designed screen wall units or grille block, concrete masonry offers a new dimension in screen wall design. The many sizes, shapes, colors, patterns, and textures available help in creating imaginative designs. Several units may make up a design, or each screen wall unit may constitute a design in itself.

Although the number of designs for concrete masonry screen units is virtually unlimited, it is advisable to check on availability of any specific unit during the early planning stage. Some designs are available only in certain localities and others are restricted by patent or copyright. A few screen units are shown in Fig. 1-17 (page 18).

Screen units should be of high quality, even though they are not often used in load-bearing construction. When tested with their hollow cells parallel to the direction of a test load, screen units should have a

Fig. 8-10. A concrete grille block wall makes a pleasing backdrop and screen.

minimum compressive strength of 1,000 psi on the gross area.

Type M or S mortar should be used for exterior screen walls and where the screen walls are required to carry any vertical load. For interior non-load-bearing walls, the mortar may be Type M, S, or N. Grout for embedding steel reinforcement in horizontal or vertical cells should comply with the criteria described in Chapter 2.

Design and Construction

Care must be taken that screen walls are stable and safe, although from a design standpoint they are seldom required to support more load than their own weight. They can be designed as load-bearing walls, but such construction is not permitted by some building codes. In any case, extra attention to design of screen walls for wind forces is warranted because of and despite the relatively high percentage of open areas. The wall openings are created by using screen units with decorative openings or, occasionally, by using conventional concrete block with intermittent vertical mortar joints, i.e. by leaving openings in the wall in lieu of vertical mortar joints.

Screen walls should be designed to resist all the horizontal forces that can be expected. Stability is provided by:

1. Using a framing system capable of carrying horizontal forces into the ground.

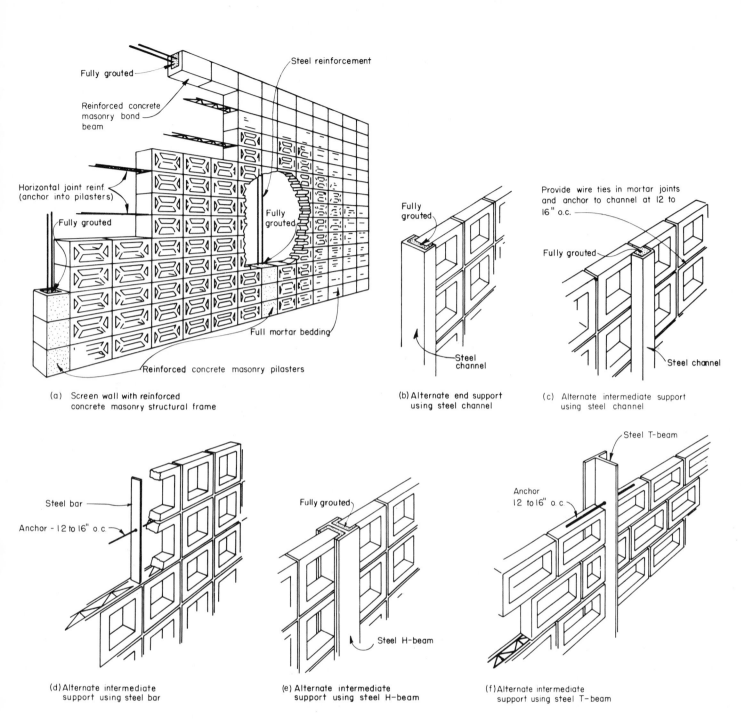

Fully grouted

Steel reinforcement

Reinforced concrete masonry bond beam

Horizontal joint reinf. (anchor into pilasters)

Fully grouted

Fully grouted

Full mortar bedding

Reinforced concrete masonry pilasters

(a) Screen wall with reinforced concrete masonry structural frame

Fully grouted

Steel channel

(b) Alternate end support using steel channel

Provide wire ties in mortar joints and anchor to channel at 12 to 16" o.c.

Fully grouted

Steel channel

(c) Alternate intermediate support using steel channel

Steel bar

Anchor - 12 to 16" o.c.

(d) Alternate intermediate support using steel bar

Fully grouted

Steel H-beam

(e) Alternate intermediate support using steel H-beam

Steel T-beam

Anchor 12 to 16" o.c.

(f) Alternate intermediate support using steel T-beam

Fig. 8-11. Framing methods for screen walls. (Adapted from TEK 5, Ref. 9.)

2. Employing adequate connection or anchorage of screen walls to the framing system.
3. Limiting the clear span.
4. Incorporating vertical reinforcement and horizontal joint reinforcement.

Partitions built with screen units are usually designed as non-load-bearing panels, with primary consideration given to adequate anchorage at panel ends and/or top edge, depending on where lateral support is furnished. Lateral design pressure is at least 5 psf.

Lateral support for screen walls may be obtained from cross walls, piers, columns, posts, or buttresses for the horizontal spans—and from floors, shelf angles, roofs, bond beams, or spandrel beams for screen walls spanning the vertical direction. The structural frame for a screen wall may consist of reinforced concrete masonry columns, pilasters, and beams, or may incorporate structural steel members as shown in Fig. 8-11. Screen wall framing methods may also be similar to those used for fences (see next section).

When designed as veneer, the concrete masonry screen wall is attached to a structural backing with wire ties or sheet metal anchors in the same manner as used for other types of masonry veneer.

A non-load-bearing screen block panel may be used to fill an opening in a load-bearing masonry wall. In this case the panel is restrained on all four sides. Joint reinforcement is placed in the horizontal joints to anchor the panel into the wall. For an exterior wall, a panel is limited to 144 sq.ft. of wall surface or 15 ft. in any direction; for an interior wall, 250 sq.ft. or 25 ft. in any direction. The lintel, sill, and jamb of the panel opening should be designed the same as for a window opening.

Non-load-bearing screen walls should have a minimum nominal thickness of 4 in. and a maximum clear span of 36 times the nominal thickness. For load-bearing screen walls, the minimum thickness should be increased to 6 in. The maximum span can be measured vertically or horizontally, but need not be limited in both directions.

Screen walls are usually capable of carrying their own weight up to 20 ft. in height, but above that height they must be supported vertically not more than every 12 ft. When screen walls support vertical loads, the allowable compressive stress should be limited to 50 psi on the gross area. In some instances the compressive stress at the base of a non-load-bearing screen wall will govern maximum unsupported height. Where

screen block are not laid in a continuous mortar bed (intermittent bond), the allowable stresses should be reduced in proportion to the reduction in the mortar-bedded area.

Due to the somewhat fragile nature of screen walls, the use of steel reinforcement is recommended wherever it can be embedded in mortar joints and bond-beam courses, or grouted into continuous vertical or horizontal cavities. When reinforced joints are used, the thickness of the mortar joint should be a minimum of twice the diameter of the reinforcement.

For exterior screen walls, joints and connections should be constructed as fully watertight as possible. The mortar joints should be made according to the best construction practices. In addition, when hollow masonry units are laid with their cores vertical, the top course should be capped to prevent the entrance of water into the wall interior.

Garden Walls and Fences

Garden walls and fences of concrete masonry can take on many delightful forms, enhancing the landscape. They are built with solid or screen block and with concrete brick or half block. If a garden wall has more than 25% open areas, it may be considered a fence. Fence framing methods are shown in Fig. 8-14.

Fences and garden walls should be able to safely

Fig. 8-12. Garden wall with regular and screen units.

Fig. 8-13. Screen block used to create a dignified property-line fence.

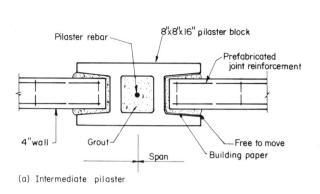

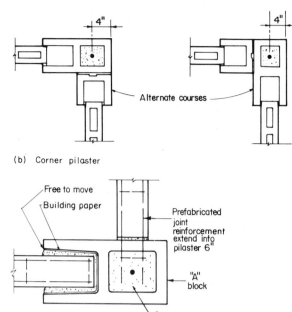

(a) Intermediate pilaster

Fig. 8-14. Framing methods for fences.

(b) Corner pilaster

(c) Corner detail

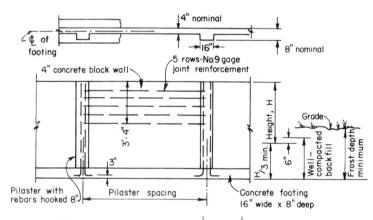

Pilaster spacing for wind pressure					Reinforcement for wind pressure			
5 psf	10 psf	15 psf	20 psf	H	5 psf	10 psf	15 psf	20 psf
19'4"	14'0"	11'4"	10'0"	4'0"	1—No. 3	1—No. 4	1—No. 5	2—No. 4
18'0"	12'8"	10'8"	9'4"	5'0"	1—No. 3	1—No. 5	2—No. 4	2—No. 5
15'4"	10'8"	8'8"	8'0"	6'0"	1—No. 4	1—No. 5	2—No. 5	2—No. 5

(a) Wall or fence with pilasters

H	Reinforcement for wind pressure			
	5 psf	10 psf	15 psf	20 psf
4'0"	1—No. 3	1—No. 3	1—No. 4	1—No. 4
5'0"	1—No. 3	1—No. 4	1—No. 5	1—No. 5
6'0"	1—No. 3	1—No. 4	1—No. 5	2—No. 4

(b) Wall or fence without pilasters

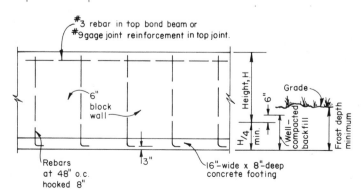

Fig. 8-15. Reinforced garden walls or fences.

withstand wind loads of at least 5 psf, and most city codes specify resistance to 20-psf pressure. Pressures and corresponding wind gust velocities are:

Pressure, psf	Wind gust velocity, mph
5	40
10	57
15	69
20	80

In hurricane areas higher wind-pressure resistance is needed.

Examples of reinforced garden walls or fences are shown in Fig. 8-15. Without reinforcement, high and straight garden walls or fences lack vertical tensile strength and are unstable in strong winds. For a 57-mph peak wind velocity the safe height of a straight 6-in. block wall is only 3 ft. 6 in.; for a straight 8-in. wall, only 5 ft. 6 in. (Fig. 6-50, page 139).

The serpentine wall or fence (Fig. 8-16) is a welcome change from the straight lines frequently seen on our modern landscapes. Undulating curves and "folded plates" give this type of wall a stability from the foundation up with no need for reinforcement.

Fig. 8-17 shows sample designs of serpentine walls based on proportions found safe for wind gusts with pressures up to 20 psf; the horizontal radius should not exceed twice the height, which in turn should not be more than twice the width. A limiting height of 15 times the thickness is recommended. The free end(s) of the serpentine wall should have additional support, such as a pilaster or a short-radius return.

Concrete masonry wall foundations may not be durable if they frequently become frozen while saturated, as noted in Chapter 2 (page 29). In cold climates, therefore, the wall foundations should be constructed with cast-in-place concrete.

Retaining Walls

Concrete masonry retaining walls can have visual beauty along with the required structural strength to resist imposed vertical and lateral loads. Because the purpose of a retaining wall is to hold back a mass of soil or other material, the design of the wall is affected by the earth's configuration—whether the earth surface behind the wall is horizontal or inclined. Design is also affected by any additional loading (surcharge), as from a vehicle or equipment passing near the top of the wall, which causes horizontal thrust on the retaining wall.

Types

There are three basic types of concrete masonry retaining walls to consider: gravity, cantilever, and counterfort or buttressed walls.

Gravity Walls

A gravity wall depends upon its own mass for stability. Basically, it is massive masonry laid so that little or no tension stress occurs in the wall under loading, and its cross-sectional shape is usually trapezoidal, as shown in Fig. 8-18. Since a retaining wall of the gravity type ordinarily has a base thickness equal to one-half to three-fourths the wall height, it is usually more economical to build a large retaining wall (more than a few feet in height) as a cantilever wall.

Fig. 8-16. A tall garden fence can be simply constructed on the serpentine principle using conventional two-core block.

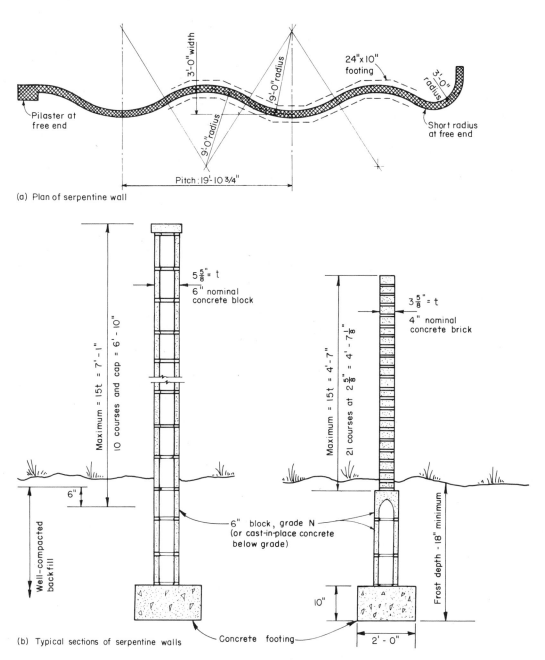

(a) Plan of serpentine wall

(b) Typical sections of serpentine walls

Fig. 8-17. Serpentine garden walls.

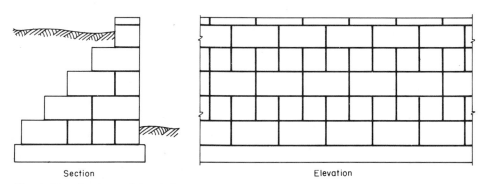

Section

Elevation

Fig. 8-18. Gravity retaining wall.

Cantilever Walls

A cantilever wall usually has a cross-sectional shape of an inverted T (Fig. 8-19), with the stem located towards the rear of the footing if soil bearing stresses are critical and towards the front or toe if sliding is critical. An L-shaped cantilever wall is used along a property line or in other situations where it is impossible to provide a toe; for such a wall, bearing pressure is usually high.

With either shape of cantilever wall, the reinforced masonry wall portion or stem performs structurally as a cantilever from the cast-in-place concrete footing. The portion of backfill directly above the footing contributes to the mass required for stability, and the concrete masonry is reinforced along the back face where the loading induces tensile stress. The functions of the footing are to hold the stem in position and to resist the forces of the stem (the sliding, overturning, and vertical pressures created by loading), transferring them to the soil.

Counterfort or Buttressed Walls

These retaining walls are similar to cantilever walls except that they span horizontally between vertical supports. Supports at the back of the wall are known as *counterforts* (Fig. 8-20), and those supports exposed at the front are called *buttresses* (Fig. 8-21).

A small degree of forward or outward tilt under service conditions is difficult to avoid with any type of retaining wall. It is therefore good practice to batter

Block width	H	a	b	t	Dowel and vertical reinforcement	Top footing reinforcement
8″	3′4″	12″	2′8″	9″	No. 3@32″ oc	No. 3@27″ oc
	4′0″	12″	3′0″	9″	No. 4@32″ oc	No. 3@27″ oc
	4′8″	12″	3′3″	10″	No. 5@32″ oc	No. 3@27″ oc
	5′4″	14″	3′8″	10″	No. 4@16″ oc	No. 4@30″ oc
	6′0″	15″	4′2″	12″	No. 6@24″ oc	No. 4@25″ oc
12″	3′4″	12″	2′8″	9″	No. 3@32″ oc	No. 3@27″ oc
	4′0″	12″	3′0″	9″	No. 3@32″ oc	No. 3@27″ oc
	4′8″	12″	3′3″	10″	No. 4@32″ oc	No. 3@27″ oc
	5′4″	14″	3′8″	10″	No. 4@24″ oc	No. 3@25″ oc
	6′0″	15″	4′2″	12″	No. 4@16″ oc	No. 4@30″ oc
	6′8″	16″	4′6″	12″	No. 6@24″ oc	No. 4@22″ oc
	7′4″	18″	4′10″	12″	No. 7@32″ oc	No. 5@26″ oc
	8′0″	20″	5′4″	12″	No. 7@24″ oc	No. 5@21″ oc
	8′8″	22″	5′10″	14″	No. 7@16″ oc	No. 6@26″ oc
	9′4″	24″	6′4″	14″	No. 8@ 8″ oc	No. 6@21″ oc

General notes:
1. Reinforcing bars should have standard deformations and a yield strength of 40,000 psi.
2. Alternate vertical reinforcing bars may be terminated at the midheight of the wall. Every third bar may be terminated at the upper third-point of the wall height.
3. The wall should have horizontal joint reinforcement at every course or else a horizontal bond beam with two No. 4 bars every 16 in.
4. Weight of assumed soil backfill (granular soil with conspicuous clay content) is 100 pcf and equivalent fluid pressure is 45 pcf. There is no surcharge and maximum soil bearing pressure is 1,500 psf.

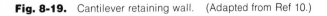

Fig. 8-19. Cantilever retaining wall. (Adapted from Ref 10.)

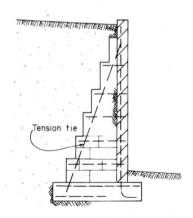

Fig. 8-20. Counterfort retaining wall.

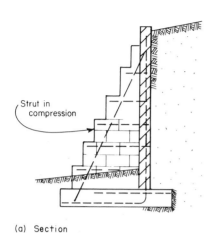
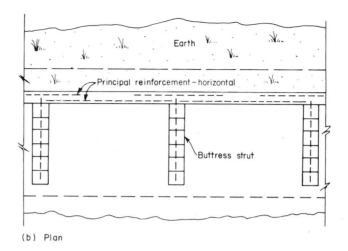

Fig. 8-21. Buttressed retaining wall.

(slope) the front face of the wall slightly to offset this tilt and avoid the illusion of instability. A batter on the order of 1/2 in. per foot is commonly used.

The selection of a particular type of retaining wall for cost and efficiency depends on the wall size, loads, soil conditions, and site location. The cantilever type of wall has a slightly lower toe pressure than the gravity type and thus may be desirable where soil bearing capacity is low. However, the gravity wall has greater resistance to sliding because of its greater weight.

It is good practice, both for design and construction of retaining walls, to use the services of an engineer who has experience with the local soil involved. He can design the cross-sectional dimensions of the wall and footing so that the computed pressure does not exceed the safe bearing value of the soil.* Table 8-3 gives safe values for different soils.

Construction

Footing

The footing for a retaining wall should be placed on firm, undisturbed soil. In areas where freezing temperatures are expected, the base of the footing is placed below the frost line. Where soil under the footing consists of soft or silty clay, 4 to 6 in. of consolidated granular fill can be placed under the footing slab to assure firm support and to increase the frictional resistance between the footing and the ground. This friction determines resistance to horizontal sliding of the wall.

Often a lug or key under the footing is provided to assist in resisting the tendency to slide (Fig. 8-22). The

*See Refs. 10 and 17 for retaining wall design examples.

Table 8-3.   Safe Bearing Pressures of Soils

Material	Bearing capacity, psf
Clay	2,000
Sand and clay mixed	4,000
Alluvium and silt	5,000
Hard clay and firm, compressed sand	8,000
Find sand	9,000
Compacted and cemented sand	10,000

same effect is achieved by requiring that the footing be well below the excavated surface in undisturbed soil, particularly if the wall is higher than 7 ft. above the footing.

Dowels to connect the wall to the footing are located so that they will be adjacent to the vertical wall reinforcement when it is placed. With a small longitudinal bar provided along the dowel line near the top of the footing, the dowels can be accurately spaced and securely tied in the correct position.

The top of the concrete footing in the area under the masonry is roughened while the concrete is still fresh. Otherwise, a 1-in.-deep, 4-in.-wide keyway is provided to improve shear bond at the joint between the wall and the footing.

Grouting and Reinforcement

The first course of block on the footing is laid in a full mortar bed. The remaining courses may then be laid with mortar coverage on the face shells and on any web between a core to be grouted and a core not to be grouted. However, there appears to be little advantage

169

in grouting only those cores containing reinforcement. If all cores are grouted, the small additional grouting material and labor costs are offset to some extent by eliminating the necessity of buttering cross webs that adjoin the cores to be grouted.

The materials and procedures previously recommended for reinforced, grouted masonry walls (Chapter 6) should be followed in the construction of retaining walls. It is necessary to provide some horizontal steel reinforcement to distribute stresses that occur when the wall expands or contracts. The amount of horizontal reinforcement needed is, to a large extent, dependent on climatic conditions. For moderate conditions and 8-in. walls, bond beams with two No. 4 bars should be placed in the top course and in intermediate courses at 16 in. on centers. For 12-in. walls, the top bond beam should contain two No. 5 bars and the intermediate bond beams should have two No. 4 bars. If desired, horizontal joint reinforcement may be placed in each joint (8 in. on center) and the bond beams omitted.

Drainage

Provisions must be made to prevent accumulation of water behind a retaining wall. Water accumulation causes increased soil pressure, seepage and, in areas subject to frost action, expansive forces of considerable magnitude near the top of the wall.

As shown in Fig. 8-22a, 4-in.-diameter weepholes spaced 5 to 10 ft. along the base of the wall should provide sufficient drainage of permeable backfill soils. An alternate weephole detail appears in Fig. 8-22c. In another alternate method, the mortar is left out of the head joints in the first or second course and about 1 cu.ft. of gravel or crushed stone is placed around the intake for each weephole.

Where unusual conditions such as heavy, prolonged rains will be encountered, seepage through weepholes may cause the ground in front of the wall and under the toe of the footing to become saturated and lose some of its bearing capacity. This undesirable condition can be avoided by installing a continuous longitudinal drain (Fig. 8-22b) that is surrounded by crushed stone or gravel near the base. Extending along the full length of the back of the wall, this drain should have outlets located beyond the ends of the wall, thus eliminating any need for weepholes. With impermeable soil and conditions tending to create excessive amounts of water in the backfill, or in areas of frequent freezing and thawing, it is advisable to provide a continuous back drain (a vertical layer of crushed stone or gravel covering the entire back of the wall) in addition to a longitudinal drain.

Other Provisions

The top of a concrete masonry retaining wall should be capped or otherwise protected to prevent entry of

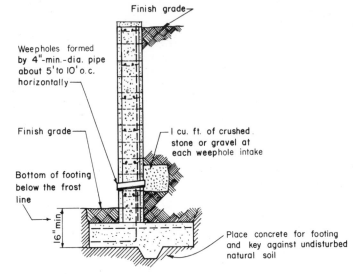

(a) With permeable backfill

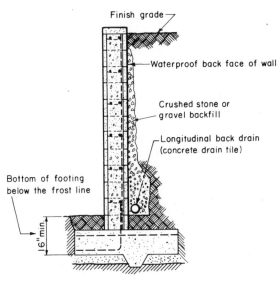

(b) With impermeable backfill

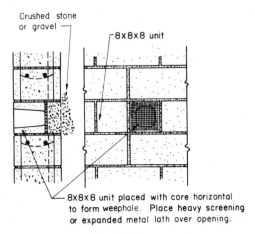

(c) Alternate weephole detail

Fig. 8-22. Suggested backfilling procedures and drainage provisions for retaining walls.

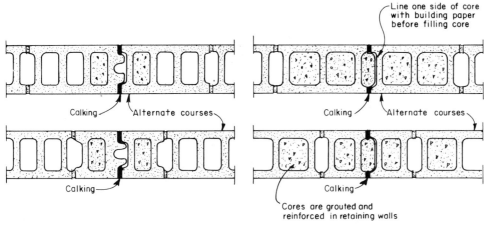

Fig. 8-23. Shear-resisting control joints for retaining walls.

water into unfilled hollow cores and spaces. Climate and type of construction will determine the need for waterproofing the back face of the wall. Since saturated mortar may not be durable in areas subject to frequent freezing and thawing, waterproofing is recommended when backfill material is relatively impermeable; it is also recommended to reduce unsightly efflorescence or leaching on the wall.

A long retaining wall should be broken into panels by means of vertical control joints, as discussed in Chapter 4. The joints should resist shear and other lateral forces in order to maintain alignment of adjacent wall sections while permitting longitudinal movement (Fig. 8-23). In some cases, to prevent seepage through a joint, it may be advisable to cover the joint with a strip of waterproofing membrane on the back of the wall.

Care should be taken when backfilling against a retaining wall. Backfilling should not be permitted until at least 7 days after grouting. It is good practice to build up the backfill material all along the wall at a rate

as nearly uniform as practicable. If heavy equipment is used in backfilling a wall designed to resist only earth pressure, such equipment should not approach the back of the wall closer than a distance equal to the height of the wall. Care should also be taken to avoid large impacting forces on the wall such as those caused by a large mass of moving earth or large stones.

Where the finished grade at the back of a retaining wall is level or nearly so, a fence or railing on top of the wall may be needed for safety. To accomplish this, for example, the masonry wall itself can be built higher by using screen block units.

Paving

Concrete masonry units are used for slope paving under highway or railway grade-separation structures and on other steep embankments to prevent costly and often dangerous soil erosion, particularly where grass will not grow to protect the surface. They are also used

Fig. 8-24.
Slope paving units are
easily laid to prevent erosion.

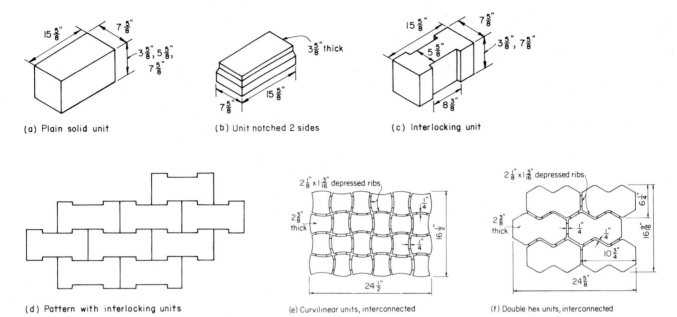

Fig. 8-25. Concrete masonry paving units.

for paving driveways, access lanes, parking areas, streets, plazas, shopping malls, walks, patios, and floors on grade, to name just a few applications. Produced in a range of shapes and colors, paving units are easy to handle and install, requiring only a few tools.

Slope Paving

With ungrouted or grouted joints, concrete masonry has been found to be an economical and pleasing solution to the slope paving problem. Construction costs are low due to minimal materials-handling at the site and the ease of placing units on the slope.

Masonry Units Required

Different sizes and shapes of paving units (Fig. 8-25) are used for slope paving, with unit thickness varying from 4 to 8 in. The thicker units are used for severe exposure such as on riverbanks. The slope paving units most often specified are standard 8×16-in. solid units* (Fig. 8-25a), although it is possible to manufacture paving units up to about 16×24 in. in size.

In areas where there are no freezing temperatures or if there is proper drainage, cored masonry units have been used successfully. They are lighter in weight and often less expensive to lay than the 100% solid units. However, the solid units are preferred in freeze-thaw climates because there is no chance of water freezing in core spaces. Also, 100% solid units discourage vandalism (by additional weight), especially in ungrouted installations.

Any concrete masonry units selected for slope paving should have a minimum compressive strength of 3,000 psi on the gross cross-sectional area at the time of delivery to the jobsite.

Construction Features

For slope paving, some specifications limit the maximum angle of slope to 35 deg.; others specify a maximum slope commensurate with the angle of repose of the underlying material. Actually, ungrouted masonry paving units can be laid on any angle at which the underlying material can be stabilized.

A 2- to 4-in. layer of granular material—sand, crushed stone, or gravel—should be installed immediately below the masonry units to facilitate drainage and minimize the possibility of frost heave. When the smoothness of the curtain of masonry units is important, sand is used because it can be struck off very smoothly. The allowable surface variation is 1/4 in. in 10 ft.

When slope paving units are not grouted together, they are laid tightly against each other. If grouting is required, 3/4- or 1-in. joints should be left for grout fill. Some contractors use wooden spacers to ensure this width, since spacing closer than 3/4 in. makes it too difficult to completely fill the joints with grout. Grout proportions, by volume, should be 1 part cement to 3 parts sand, with just enough water to make the grout

---

*Solid units for paving have no voids (100% solid) whereas solid units for other applications may have up to 25% voids.

workable. Interlocking units (Fig. 8-25) do not require grouted joints.

An ungrouted installation offers several advantages:

1. If there is settlement or frost heave, the paving units can be adjusted individually.
2. If appearance allows, the underlayment need not be so carefully struck off and thus the use of crushed stone or gravel becomes more feasible.
3. Ungrouted construction is less expensive and the units can be easily replaced.

The advantages of grouted construction are

1. Percolation of water between and under paving units is avoided.
2. There is less chance of undermining the slope protection.
3. The units are more securely held in place, deterring vandalism and theft.

Most concrete masonry slope paving installations are built with some type of support at the bottom or toe of the slope to prevent sliding of the units and to provide a straight, firm foundation for the first course of masonry. In some cases the support may be provided by a compact, level surface of embankment material. Toe construction will vary, depending on conditions of drainage, paving, and grade at the bottom of the slope (Fig. 8-26).

It should be noted that, instead of a large support at the bottom of the slope, smaller intermittent supports may be made on the slope itself. For example, each tenth course of masonry could be embedded into the slope, with the long dimension of the block perpendicular to the slope.

Drain troughs are often provided at the sides of the area paved with masonry (Fig. 8-27) and channelized waterways also carry water from the toe of the slope to a natural outlet. These waterways should be paved for a sufficient distance to eliminate any possibility of erosion undermining the slopes.

Where drains are not necessary, care should be exercised that the fill material at the edge of the paving units is level with the top of the units. This is especially important to hold the edge units in place when joints are not grouted.

Riverbank Revetment

As an alternative to stone riprap, special concrete revetment block are available for controlling erosion along the banks of rivers and lakes. The perforated waffle units shown in Fig. 8-28a may be laid on the bank beneath the water, one unit adjacent to the next. Grout is needed only on the perimeter of the revetment to keep these units in place, and construction costs can be lowered by inserting precast keys or stakes in the preformed key holes. The cores in the units may be filled with sand and gravel, and vegetation will grow through the cores, further securing the

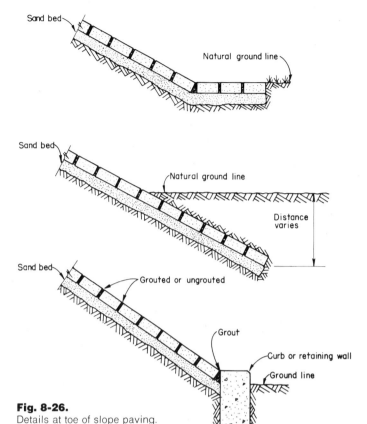

Fig. 8-26.
Details at toe of slope paving.

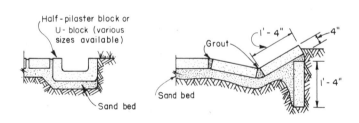

Fig. 8-27. Edge drains for slope paving.

units in position. Such a revetment functions as an articulated mat, with each unit settling individually to a firm resting position.

The ribbed waffle units (Fig. 8-28b) used for erosion control are 90-lb. units that can be laid by unskilled volunteers during flood emergencies. Interlocking units (Fig. 8-25c) are also used for erosion control.

Other Paving

Any of the paving units in Fig. 8-25 can be used for foot and light vehicle traffic. Furthermore, the concrete masonry industry offers a variety of interlocking

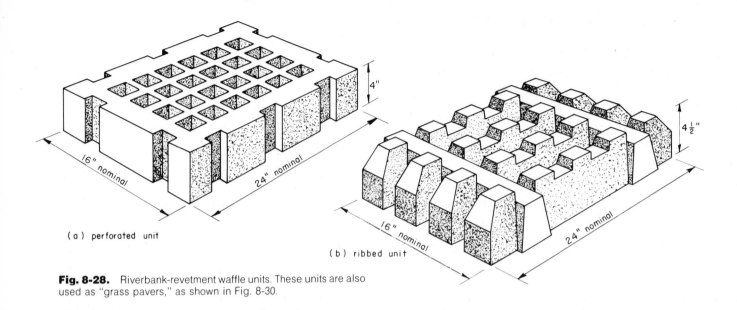

(a) perforated unit

(b) ribbed unit

Fig. 8-28. Riverbank-revetment waffle units. These units are also used as "grass pavers," as shown in Fig. 8-30.

Fig. 8-29. Interlocking concrete masonry pavers make an uncommonly attractive auto showroom floor.

Fig. 8-30. Ribbed waffle units may be used for overflow parking areas. Grass grows between the ribs to enhance appearance.

paving units in interesting patterns, bright with color and human in scale. A popular interlocking paver is pictured in Fig. 8-29.

Waffle units (Fig. 8-28) are useful as turf block or "grass pavers," as pictured in Fig. 8-30. They are laid to create light pavement at access lanes and parking areas. Grass will grow up in the cores or between the ribs despite frequent traffic or parking of vehicles.

Patio block are commonly 8×16 in. (nominal), with thicknesses of 1-5/8, 2-1/4, and 3-5/8 in. Various shaped units are available in a natural grey, white (by use of white cement), or tones of red, black, brown, green, or yellow. Almost any pattern used for masonry walls (Fig. 4-16, pages 82-88) may be used for patio block.

Installation of paving units can begin in the spring as soon as the frost is gone and the ground is dry enough to work on. In the fall, installation can continue until frozen ground prevents proper compaction of the subgrade material. The final job will be only as stable as the subgrade.

Requiring careful preparation, the subgrade should be uniform, hard, free from foreign matter, and well drained. The best masonry paving installations are made by removing all organic matter such as grass, sod, and roots. Hard spots are loosened and tamped to provide the same uniform support as the rest of the subgrade. Mucky spots are dug out, filled with soil similar to the rest of the subgrade or with granular material (such as sand, gravel, crushed stone, or slag), and compacted thoroughly. All fill materials should be uniform, free of vegetable matter, large lumps, large stones, and frozen soil.

Paving units are bedded either in sand or in a durable air-entrained mortar. Sand bedding can give satisfactory, long-lasting results and is ideally suited for the do-it-yourselfer in such applications as walks, patios, and temporary pathways. The units are easily lifted for maintenance work or relocated if desired. Edge restraint such as concrete masonry curbs may be needed to prevent the units from creeping apart.

If mortar bedding is used, the surface of the paving units must be provided with positive drainage.

Catch Basins and Manholes

Concrete masonry construction is an accepted method for building catch basins, inlets, manholes, valve vaults, pump wells, and other shallow-depth, circular, underground structures. A typical catch basin and some sewer manholes are shown in Figs. 8-31 and 8-32, respectively.

Catch basins have space at the bottom for the settlement and storage of suspended solids that might otherwise be carried and deposited in the pipeline. If the catch basins are part of a sanitary sewerage system, they should be provided with solid covers to prevent sewer odors from reaching the street.

Drop manholes are constructed at intersecting lines or where there is an abrupt drop in elevation in a sewer line. The arrangement shown in Fig. 8-32c is desirable in that it reduces turbulence and prevents sewage from splashing on men working in the manhole. Such construction still permits rodding out and lamping the sewer.

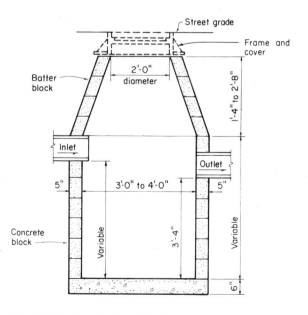

Fig. 8-31. A typical catch basin.

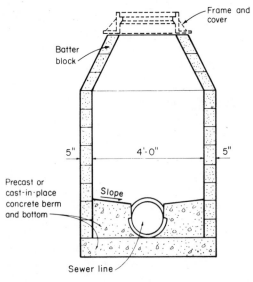

(a) Section of manhole for sewers 33" or less in dia.

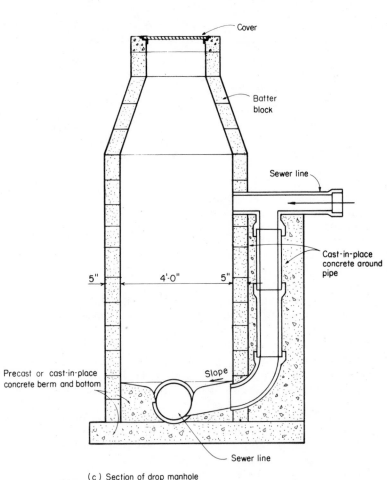

(c) Section of drop manhole

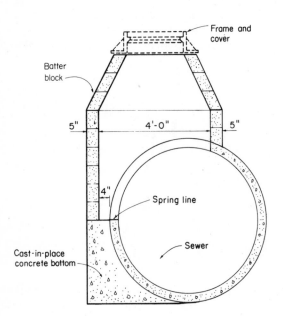

(b) Section of manhole for large sewers more than 33" in dia.

Fig. 8-32. Types of sewer manholes.

Materials

Concrete block should meet the absorption and strength requirements of Standard Specification for Concrete Masonry Units for Construction of Catch Basins and Manholes, ASTM C139. This specification is for segmental masonry units. When concrete brick are used, they should be plastered on the outside with Type S mortar.

Batter block (Fig. 1-16, page 17) are very useful for cone construction atop the barrel of a catch basin (Fig. 8-33) or manhole. The cone reduces the inside diameter to 2 ft. at the top to receive a standard manhole cover, as shown in Figs. 8-31 and 8-32. No cutting of units is necessary because block producers have predetermined the exact number and size of batter block required.

For a manhole cover size for which batter block are not available, the masonry wall of the structure is continued without reduction for its entire height. A precast or cast-in-place concrete slab is then added.

Special block are available to frame around inlet and outlet pipe.

Construction

For a catch basin or manhole, excavation should be to the required depth and dimensions specified. If rock is encountered, the bottom of the excavation should be carried down at least 6 in. below the elevation of the bottom of the structure and backfilled with sand.

The bottom of the structure may be constructed of cast-in-place concrete or a precast concrete slab, with the concrete having a water-cement ratio, by weight, of no more than 0.50 and a minimum compressive strength of 3,000 psi at 28 days. The use of a precast bottom is finding increasing favor because it minimizes

the time the excavation must be kept open. It is of particular value during wet weather or when the excavation is in wet subsoil.

The masonry wall of the structure is constructed in horizontal courses, with vertical joints staggered. All joints are completely filled with Type M mortar. Any castings that will be used should be set in a full bed of mortar. The masonry units around inlet or outlet pipe are carefully laid and sealed with mortar to prevent leakage. Special block are useful to frame around inlet and outlet pipe.

Heavily galvanized or other noncorrosive ladder rungs may be attached to a manhole wall or embedded in it on about 16-in. centers. Otherwise, a ladder may be installed in the manhole.

Granular material such as sand, gravel, or crushed stone is used to backfill the completed structure. Backfill material may be governed by specifications or approval of the engineer.

Storage Bins

Concrete masonry units are popular for construction of storage bins for grains, fruits, and vegetables. Where drying of the stored produce is not important, masonry units are generally laid in the conventional manner. Where produce drying is necessary, the units are laid with cores horizontal, as shown in Fig. 8-34.

Details of construction for concrete masonry storage bins are given in Fig. 8-35. Pilasters for lateral support against the pressure of the stored produce may be built of reinforced concrete masonry or reinforced concrete. They may be placed outside the walls (Fig. 8-35) or flush (Fig. 8-34).

Protection against rodents is provided by installation of hardware cloth and metal joint strips, as shown in Fig. 8-36.

Fig. 8-33. Concrete block catch basin in a storm sewer system.

Fig. 8-34. Built of concrete block laid with cores horizontal, this rectangular crib provides durable, safe storage for ear corn.

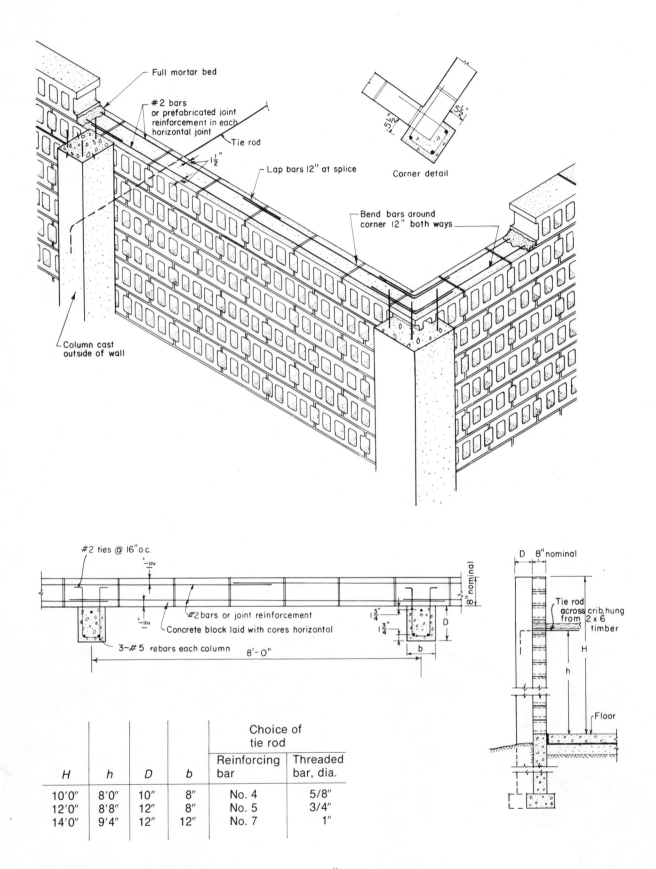

Fig. 8-35. Details of a rectangular concrete masonry corn crib.

H	h	D	b	Choice of tie rod	
				Reinforcing bar	Threaded bar, dia.
10'0"	8'0"	10"	8"	No. 4	5/8"
12'0"	8'8"	12"	8"	No. 5	3/4"
14'0"	9'4"	12"	12"	No. 7	1"

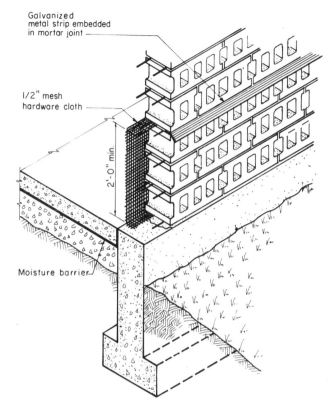

Galvanized metal strip embedded in mortar joint

1/2" mesh hardware cloth

2'-0" min.

Moisture barrier

Fig. 8-36. Installation of metal strip shield and hardware cloth to keep rodents out of a concrete masonry corn crib.

Floors and Roofs

Accompanying the continuously expanded use of concrete masonry units for walls and partitions of buildings has been a steady interest in using concrete masonry units for floor and roof construction. Concrete masonry floors and roofs are well suited to all types of structures except those involving very heavy loadings.

The original use of concrete masonry for floor and roof construction began many years ago with the application of filler units between cast-in-place concrete joists. The more recent development of industrialized systems or methods of construction has encouraged the further refinement of floor and roof construction with masonry units. The newer systems have resulted in decreased costs and reduced expenditure of jobsite time as well as improved structural performance; i.e., greater load-carrying capacity and smaller dead-load deflections. In most of these systems a

cast-in-place topping is required to complete the construction.

Systems Commonly Used

Filler or Soffit Block

The type of system for floor and roof construction shown in Fig. 8-37 combines concrete masonry—filler or soffit block—with cast-in-place concrete joists and slab. The concrete block are placed on the formwork in rows, with spaces left between the rows to form the joists. Reinforcement is then placed in these spaces before the concrete for joists and slab is placed. The concrete bonds to and holds the block in position.

Filler or soffit block not only provide the forms for concrete placement but also form a temporary deck for workmen installing mechanical, electrical, and communications needs. The block may be easily cut or drilled to provide access space for wiring and other services.

A filler or soffit block system can use masonry units of any thickness. The slab over the joists can also be of any desired thickness. Therefore, such construction is applicable to floors and roofs having a correspondingly wide range of effective depths, maximum span limits, and load capacities. Structural design by the ACI 318 Building Code allows shear and bending stresses to be figured on the outside shell of the block.

Block with Composite Joists

This type of system (Figs. 8-38 and 8-39) uses block laid dry between precast concrete or other joist members, all made monolithic by site-cast concrete. In some of the patented systems available, steel joists combine a bottom chord encased in concrete with an exposed top webbing; these joists are used with wide-flange filler block. Other systems are inverted T-section concrete joists. Temporary shoring of joists may be necessary to ensure correct camber. Spans usually range up to about 30 ft., depending on block depth.

Prefabricated Block Plank

Block plank are plant-prefabricated by threading together concrete masonry units, normal reinforcing bars, and prestressing strands (Fig. 8-40). Contact surfaces between block are ground to assure proper bearing, and tongue-and-groove units are used for easy mating of adjoining plank when forming a floor or roof. Mild steel bars may be inserted in the small upper holes of the units for development of negative moments that occur at the support of cantilevers. The reinforcing steel and prestressing strands are bonded by injecting grout into the voids. A site-cast concrete topping may be used, and spans up to 33 ft. are possible.

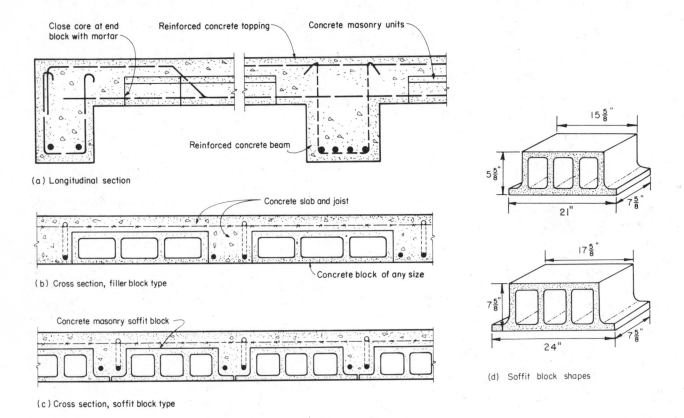

(a) Longitudinal section

(b) Cross section, filler block type

(c) Cross section, soffit block type

(d) Soffit block shapes

Fig. 8-37. Filler or soffit block construction of floors and roofs.

Fig. 8-38.
Concrete masonry units are set
between prefabricated joists and then concrete
is cast over and between them.

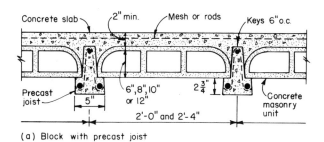

(a) Block with precast joist

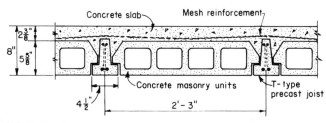

(b) Block with precast joist

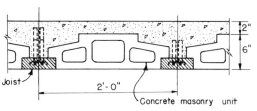

NRC is approximately 0.45 and STC is greater than 40

(c) Block with composite joist

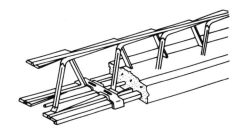

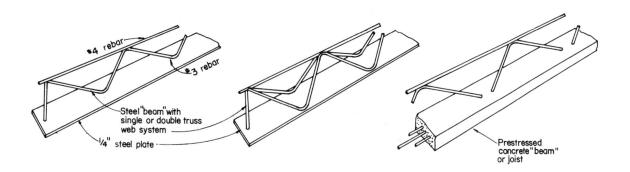

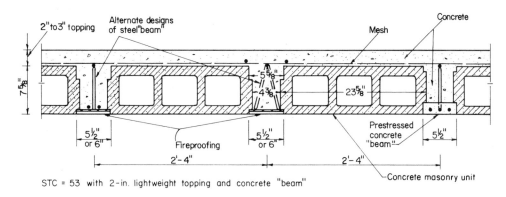

STC = 53 with 2-in. lightweight topping and concrete "beam"

(d) Block with composite joist or steel "beam"

Fig. 8-39. Block and precast joist construction.

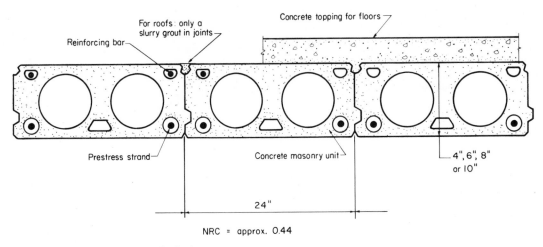

For roofs: only a slurry grout in joints

Concrete topping for floors

Reinforcing bar

Prestress strand

Concrete masonry unit

4", 6", 8" or 10"

24"

NRC = approx. 0.44

Fig. 8-40. Prefabricated block plank.

Construction

With all these floor and roof systems, construction techniques are relatively simple and no special equipment is required. Other advantages include a reduced dead load, increased fire resistance, improved acoustics and thermal insulation, integral chaseway for pipe and wire, and in most cases a flat soffit for a ceiling. When these factors are taken into consideration, concrete masonry floors and roofs can be quite economical.

Good bond between masonry units and any concrete topping is necessary. It is, therefore, essential that the masonry surface be free of debris, dirt, oil, and other construction-job litter. A workable concrete mixture with a 3- to 5-in. slump should be used for the topping, making certain that the concrete is vibrated around the reinforcement and into intimate contact with all masonry surfaces.

All of the floor and roof systems discussed are characterized by elimination or significant reduction of formwork. Although shoring is usually needed, it is much simplified from that generally required. If the system uses cast-in-place joists, their importance to the structural performance of the system requires that shoring and forming remain in place until the freshly cast concrete has attained sufficient strength to support its own weight. The engineer or architect should specify when the forms and shores can be stripped and the criteria for reshoring.

Deflections should be considered when the floor and roof system is selected. Systems using normal reinforcement are subject to downward deflections, while those with prestressed members are sometimes subject to upward deflections or "camber growth." These deflections can be computed and proper allowance made; e.g., by cambering the forms.

Lightweight concrete masonry units are sometimes used in floor and roof construction to reduce the dead weight of the structure. Also, these units have excellent acoustical and thermal properties desired by architects.

Prefabricated Panels

Prefabrication in the construction industry extends to concrete masonry wall panels built away from the building for speed, efficiency, and quality of construction. In one elementary method, concrete masonry wall panels are constructed in a plant over reinforced concrete grade beams and then trucked to the site and set in place with a crane (Fig. 8-41). In another method the concrete masonry wall panel is constructed on a working slab. Lifting hooks are built into the panel, and the strength of the grout holds the panel together when it is lifted by the top. If the panel is to be moved to a storage area for curing, a steel bed frame is used for support.

Wall panels built by hand (Fig. 8-43) may use mass-production refinements. For example, mortar proportioning and mixing can be controlled carefully, and special mortar spreaders (Fig. 6-3, page 123) have been developed to lay the mortar on face shells without the usual spilling and waste. Also, electric scaffolds that rise with the work permit masons to be always in the most comfortable and productive position.

Wall panels may also be built offsite by a machine that lays block at the rate of one every five seconds, much faster than a mason can work. The machine completely fills the head joints as the concrete block is vibrated into place. Cross webs as well as face shells are mortared using this highly mechanized operation.

Fig. 8-41. Panel of concrete masonry preassembled for residential construction.

Fig. 8-42. Prefabricated wall panels are braced in position until the framing above is completed.

Fig. 8-43. Masonry panels are fabricated at the site in a canvas-covered shelter, where they are stored for 7 days before being lifted by crane onto the building.

The mortar is of very high quality; no admixtures are used.

The cost of prefabricated wall panels is comparable to the cost of walls that are laid in place. However, there are advantages of prefabrication that produce appreciable savings and other benefits. Not only are better quality and uniformity of construction more easily achieved, but also the speed of erection is quite distinct. Work can proceed on the panels in bad weather. Masons working in shop buildings or other enclosures are assured continuous work with little interruption. Also, cumbersome scaffolding and its winter enclosures are eliminated, and the accident rate among masons in prefabrication work is lower.

Fig. 8-44.
Plant fabrication of panels. Concrete block are conveyed by belt to a machine that automatically assembles them into wall panels. The finished panels are rolled along a track to a storage area.

Since this handbook points out only the most essential facts concerning concrete masonry and its constituent materials, references have been made to many of the following publications to indicate the sources of material presented in the text. This list of references, grouped here into various classifications, is also intended to serve as a guide for further study.

General References

1. *Masonry: Past and Present,* STP 589, American Society for Testing and Materials, Philadelphia, Pa., 1975.

2. *Menzel Symposium on High Pressure Steam Curing,* SP-32, American Concrete Institute, Detroit, Mich., 1970.

3. Menzel, Carl A., *General Considerations of Cracking in Concrete Masonry Walls and Means for Minimizing It,* Bulletin D20, Research and Development Laboratories, Portland Cement Association, Skokie, Ill., 1958.

4. Copeland, R. E., and Carlson, C. C., "Tests of the Resistance to Rain Penetration of Walls Built of Masonry and Concrete," *Journal of the American Concrete Institute,* Proceedings Vol. 36, November 1939, pages 169-192.

5. Hedstrom, R. O., *Load Tests of Patterned Masonry Walls,* Bulletin D41, Research and Development Laboratories, Portland Cement Association, Skokie, Ill., 1961.

6. Hedstrom, R. O., *Tensile Testing of Concrete Block and Wall Elements,* Bulletin D105, Research and Development Laboratories, Portland Cement Association, Skokie, Ill., 1966.

7. Kuenning, W. H., *Improved Method of Testing Tensile Bond Strength of Masonry Mortars,* Research Bulletin 195, Research and Development Laboratories, Portland Cement Association, Skokie, Ill., 1966.

8. Hedstrom, R. O.; Litvin, Albert; and Hanson, J. A., *Influence of Mortar and Block Properties on Shrinkage Cracking of Masonry Walls,* Bulletin D131, Research and Development Laboratories, Portland Cement Association, Skokie, Ill., 1968.

9. *TEK Information Series,* National Concrete Masonry Association, McLean, Va., continuing series from 1970.

10. *Reinforced Concrete Masonry Retaining Walls,* National Concrete Masonry Association, McLean, Va., 1954.

11. Northwood, T. D., and Monk, D. W., *Sound Transmission Loss of Masonry Walls: Twelve-Inch Lightweight Concrete Blocks with Various Surface Finishes,* Building Research Note No. 90, National Research Council of Canada, Ottawa, Ont., Canada, April 1974.

12. *Nonreinforced Concrete Masonry Design Tables,* National Concrete Masonry Association, McLean, Va., 1971.

13. Amrhein, James E., *Masonry Design Manual,* Masonry Institute of America (formerly Masonry Research), Los Angeles, Calif., 1969.

14. Amrhein, James E., *Reinforced Masonry Engineering Handbook,* Second Edition, Masonry Institute of America (formerly Masonry Research), Los Angeles, Calif., 1973.

15. *Residential Fireplace and Chimney Handbook,* Masonry Institute of America (formerly Masonry Research), Los Angeles, Calif., 1970.

16. *All-Weather Masonry Construction—State-of-the-Art Report,* Technical Task Committee report, International Masonry Industry All-Weather Committee, December 1968.

17. Sears, Bradford G., *Retaining Walls,* American Society of Landscape Architects Foundation, McLean, Va., 1973.

18. *Plasterer's Manual,* EB049M, Portland Cement Association, Skokie, Ill., 1962.

19. Fishburn, Cyrus C., *Strength and Resistance to Corrosion of Ties for Cavity Walls,* Report BMS 101, National Bureau of Standards, U.S. Department of Commerce, Washington, D.C., July 1943.

20. *Reinforced Concrete Masonry Design Tables,* National Concrete Masonry Association, McLean, Va., 1971.

21. Allen, L. W.; Stanzak, W. W.; and Galbreath, M., *Fire Endurance Tests on Unit Masonry Walls with Gypsum Wallboard,* NRCC No. 13901, National Research Council of Canada, Ottawa, Ont., Canada, February 1974.

22. *Design Manual, Plain Concrete Block Masonry Bearing Walls in High Rise Buildings,* National Concrete Producers Association, Downsview, Ont., Canada, 1969.

23. *Design Guide to Concrete Masonry Multistorey Buildings,* National Concrete Producers Association, Downsview, Ont., Canada, April 1973.

24. Leba, Theodore, Jr., *Design Manual, The Application of Non-Reinforced Concrete Masonry Load-Bearing Walls in Multi-Storied Structures,* National Concrete Masonry Association, McLean, Va., 1969.

25. Mackintosh, Albyn, *Design Manual, The Application of Reinforced Concrete Masonry Load-Bearing Walls in Multi-Storied Structures,* National Concrete Masonry Association, McLean, Va., 1968.

26. Isberner, Albert W., Jr., *Specifications and Selection of Materials for Masonry Mortars and Grouts,* Bulletin RD024M, Research and Development Laboratories, Portland Cement Association, Skokie, Ill., 1974.

27. Isberner, Albert W., Jr., *Properties of Masonry Cement Mortars,* Bulletin RD019M, Research and Development Laboratories, Portland Cement Association, Skokie, Ill., 1974.

28. *How to Calculate Heat Transmission Coefficients and Vapor Condensation Temperatures of Concrete Masonry Walls,* IS015M, Portland Cement Association, Skokie, Ill., 1946.

29. *ASHRAE Handbook of Fundamentals,* American Society of Heating, Refrigerating and Air-Conditioning Engineers, Inc., New York, N.Y., 1972.

30. Brewer, Harold W., *General Relation of Heat Flow Factors to the Unit Weight of Concrete,* Bulletin D114, Research and Development Laboratories, Portland Cement Association, Skokie, Ill., 1967.

31. Peavy, B. A.; Powell, F. J.; and Burch, D. M., *Dynamic Thermal Performance of an Experimental Masonry Building,* Building Science Series No. 45, National Bureau of Standards, U.S. Department of Commerce, Washington, D.C., 1973.

32. *The Concrete Approach to Energy Conservation,* SP014B, Portland Cement Association, Skokie, Ill., 1974.

33. Harris, C. M., *Handbook of Noise Control,* McGraw-Hill Book Co., New York, N.Y., 1957.

34. Randall, Frank A., Jr., *Acoustics of Concrete in Buildings,* TA037B, Portland Cement Association, Skokie, Ill., 1974.

35. *Noise Abatement and Control,* Highway Research Record No. 448, Transportation Research Board (formerly the Highway Research Board), Washington, D.C., 1973.

36. Menzel, C. A., *Tests of the Fire Resistance and Strength of Walls of Concrete Masonry Units,* Portland Cement Association, Skokie, Ill., January 1934.

37. Catani, Mario J., and Goodwin, Stanley E., "Heavy Building Envelopes and Dynamic Thermal Response," *Journal of the American Concrete Institute,* Proceedings Vol. 73, February 1976, pages 83-86.

38. Allen, L. W., *Fire Endurance of Selected Non-Loadbearing Concrete Masonry Walls,* Fire Study No. 25, Division of Building Research, National Research Council of Canada, Ottawa, Ont., Canada, March 1970.

39. *Building Materials List,* Underwriters' Laboratories, Inc., Chicago, Ill., January 1974, pages 440-442.

40. Yokel, F. Y.; Mathey, R. G.; and Dikkers, R. D., *Compressive Strength of Slender Concrete Masonry Walls,* Building Science Series No. 33, National Bureau of Standards, U.S. Department of Commerce, Washington, D.C., 1970.

41. Schmidt, J. L.; Olin, H. B.; and Lewis, W. H., *Construction Principles, Materials & Methods,* Second Edition, U.S. Savings and Loan League, Chicago, Ill., 1972.

42. Yokel, Felix Y., "Engineers Inquiry Box," *Civil Engineering-ASCE,* December 1974, page 60.

43. *Efflorescence,* IS020T, Portland Cement Association, Skokie, Ill., 1968.

44. Holm, Thomas A., "Engineered Masonry with High Strength Lightweight Concrete Masonry Units," *Concrete Facts,* Vol. 17, No. 2, Expanded Shale, Clay and Slate Institute, Washington, D.C., 1972, pages 9-16.

45. Litvin, Albert, *Clear Coatings for Exposed Architectural Concrete,* Bulletin D137, Research and Development Laboratories, Portland Cement Association, Skokie, Ill., 1968.

46. Dickey, Walter L., "Concrete Masonry Construction," *Handbook of Concrete Engineering,* ed. Mark Fintel, Van Nostrand Reinhold Company, New York, N.Y., 1974, pages 536-571.

47. *Fire Resistance Index,* Underwriters' Laboratories, Inc., Chicago, Ill., January 1975.

48. *Fire Resistance of Expanded Shale, Clay and Slate Concrete Masonry,* Lightweight Concrete Information Sheet No. 14, Expanded Shale, Clay and Slate Institute, Washington, D.C., October 1971.

49. *Thermal Insulation of Various Walls,* Expanded Shale, Clay and Slate Institute, Washington, D.C., July 1972.

50. Diehl, John R., *Manual of Lathing and Plastering,* Mac Publishers Association, New York, N.Y., 1960 (out of print).

51. Portland Cement Association, *Special Concretes, Mortars, and Products,* John Wiley & Sons, Inc., New York, N.Y., 1975, pages 269-342.

Codes, Regulations, and Recommended Practices

52. ACI Committee 531, "Concrete Masonry Structures—Design and Construction," *Journal of the American Concrete Institute,* Proceedings Vol. 67, No. 5, May 1970, pages 380-403, and No. 6, June 1970, pages 442-460.

53. ACI Committee 531, "Proposed Standard Specifications for Concrete Masonry," *Journal of the American Concrete Institute,* Proceedings Vol. 72, No. 11, November 1975, pages 614-627.

54. *A Guide for the Design and Construction of Unit Masonry (A224),* Canadian Standards Association, Rexdale, Ont., Canada, 1970.

55. "Plain and Reinforced Masonry," *Commentaries on Part 4, Supplement No. 4 to the National Building Code of Canada,* NRCC No. 13989, National Research Council of Canada, Ottawa, Ont., Canada, 1975, pages 1-41.

56. *Building Code Requirements for Reinforced Masonry (A41.2),* National Bureau of Standards Handbook 74, American National Standards Institute, New York, N.Y., 1960.

57. *American Standard Building Code Requirements for Masonry (A41.1),* National Bureau of Standards Misc. Publication 211, American National Standards Institute, New York, N.Y., 1954.

58. *Specification for the Design and Construction of Load-Bearing Concrete Masonry,* TR75-B, National Concrete Masonry Association, McLean, Va., 1968.

59. *Masonry Structural Design for Buildings,* TM5-809-3, Department of the Army, or AFM 88-3, Chapter 3, Department of the Air Force, Washington, D.C., 1973.

60. *Seismic Design for Buildings,* TM5-809-10, Department of the Army, or AFM 88-3, Chapter 13, Department of the Air Force, or NAVDOCKS P-355, Department of the Navy, Washington, D.C., 1966.

61. *Fire-Resistance Classifications of Building Construction,* Report BMS 92, National Bureau of Standards, U.S. Department of Commerce, Washington, D.C., 1942.

62. *Recommended Practices & Guide Specifications for Cold Weather Masonry Construction,* International Masonry Industry All-Weather Council, 1970. Available from Portland Cement Association, Skokie, Ill., as LT107M or from other Council members (listed in Table 5-1 footnote, page 112).

63. *Guide Specification for Military and Civil Works Construction—Masonry,* CE-206.01, Corps of Engineers, Department of the Army, Washington, D.C., 1968.

64. *Manual of Acceptable Practices to the HUD Minimum Property Standards,* 1973 Edition, Vol. 4, U.S. Department of Housing and Urban Development, Washington, D.C., 1973.

65. *Guide Specification for Concrete Masonry,* National Concrete Masonry Association, McLean, Va., 1971.

66. *Concrete Masonry Units,* Seventh Edition, UL618, Underwriters' Laboratories, Inc., Chicago, Ill., 1975.

67. *Minimum Property Standards for One- and Two-Family Dwellings,* 1973 Edition, Vol. 1, U.S. Department of Housing and Urban Development, Washington, D.C., 1973.

68. *Fire-Performance Ratings 1975,* Supplement No. 2 of the National Building Code of Canada, NRCC No. 13987, National Research Council, Ottawa, Ont., Canada, 1975.

69. *Fire Protection Handbook,* 13th Edition, National Fire Protection Association, Boston, Mass., 1969.

70. *Recommended Practice for Atmospheric Pressure Steam Curing of Concrete (ACI 517-70),* American Concrete Institute, Detroit, Mich., 1970.

Specifications and Methods of Testing

American Society for Testing and Materials (ASTM)

C5	Standard Specification for Quicklime for Structural Purposes
C31	Standard Method of Making and Curing Concrete Test Specimens in the Field
C33	Standard Specification for Concrete Aggregates
C39	Standard Method of Test for Compressive Strength of Cylindrical Concrete Specimens
C55	Standard Specification for Concrete Building Brick
C90	Standard Specification for Hollow Load-Bearing Concrete Masonry Units
C91	Standard Specification for Masonry Cement
C94	Standard Specification for Ready-Mixed Concrete
C129	Standard Specification for Non-Load-Bearing Concrete Masonry Units
C139	Standard Specification for Concrete Masonry Units for Construction of Catch Basins and Manholes
C140	Standard Methods of Sampling and Testing Concrete Masonry Units
C143	Standard Method of Test for Slump of Portland Cement Concrete
C144	Standard Specification for Aggregate for Masonry Mortar
C145	Standard Specification for Solid Load-Bearing Concrete Masonry Units
C150	Standard Specification for Portland Cement
C207	Standard Specification for Hydrated Lime for Masonry Purposes
C236	Test for Thermal Conductance and Transmittance of Built-up Sections by Means of the Guarded Hot Box
C270	Standard Specification for Mortar for Unit Masonry
C315	Standard Specification for Clay Flue Linings
C404	Standard Specification for Aggregates for Masonry Grout
C426	Standard Method of Test for Drying Shrinkage of Concrete Block
C476	Standard Specification for Mortar and Grout for Reinforced Masonry
C595	Standard Specification for Blended Hydraulic Cements
C617	Standard Method of Capping Cylindrical Concrete Specimens
C744	Standard Specification for Prefaced Concrete and Calcium Silicate Masonry Units
C780	Standard Method for Pre-Construction and Construction Evaluation of Mortars for Plain and Reinforced Masonry
E72	Standard Methods of Conducting Strength Tests of Panels for Building Construction
E119	Methods of Fire Tests of Building Construction and Materials
E380	Standard Metric Practice Guide (A Guide to the Use of SI—the International System of Units)

Canadian Standards Association (CSA)

A5	Portland Cements
A8	Masonry Cement
A23.1	Concrete Materials and Methods of Concrete Construction
A82.30	Interior Furring, Lathing and Gypsum Plastering
A82.42	Quicklime for Structural Purposes
A82.43	Hydrated Lime for Masonry Purposes
A82.56	Aggregate for Masonry Mortar
A165.1	Concrete Masonry Units
A165.2	Concrete Brick Masonry Units
A165.3	Prefaced Concrete Masonry Units

A165.4 Coreless Autoclaved Cellular Concrete Masonry Units for Load-Bearing and Non-Load-Bearing Use

A179 Mortar for Unit Masonry

B54.3 Methods of Fire Tests of Walls, Partitions, Floors, Roofs, Ceilings, Columns, Beams, and Girders

American National Standards Institute (ANSI)

A10.9 American National Standard Safety Requirements for Concrete Construction and Masonry Work

A42.2 Standard Specifications for Portland Cement and Portland Cement-Lime Plastering, Exterior (Stucco) and Interior

A42.3 Standard Specifications for Lathing and Furring for Portland Cement and Portland Cement-Lime Plastering, Exterior (Stucco) and Interior

A58.1 Building Code Requirements for Minimum Design Loads in Buildings and Other Structures

DETAILS OF CONCRETE MASONRY CONSTRUCTION

On the following pages are a number of details that might be encountered in buildings using concrete masonry. These details are offered only as a *guide* for design and are not guaranteed for completeness or suitability for all buildings in all places. Building designs involve a wide range of shapes, dimensions, materials, loads, uses, and climates. While some simple details are adequate for many ordinary structures, there are circumstances that require they be refined or improved, as in regions with earthquakes or high winds. The knowledgeable and experienced designer or builder will recognize the applications of various configurations and be able to refine or improve these details to fit the individual project.

The connection of floors and roofs to masonry walls deserves particular attention. Some building codes, particularly where earthquakes are common, require a positive connection between the floor or roof and the masonry wall.* Although practice varies from one region to another, engineered concrete masonry structures most often have reinforcing bar connections between wall and floor or roof. Wooden framing must always be anchored. In some cases the dead-load friction of a concrete or steel deck on the wall will suffice. The following friction coefficients, which are

based on a safety factor of 2, are suggested:

Connection	Friction coefficient
Steel to steel	0.12
Cast-in-place concrete to steel	0.20
Cast-in-place concrete to hardboard	0.25
Cast-in-place concrete to cast-in-place concrete	0.40
Precast concrete to concrete masonry	0.40
Cast-in-place concrete to concrete masonry	0.50

Where a floor is embedded in a wall, the connection will be stronger than for a roof that is merely resting on top of the wall. A roof may be subject to temperature and shrinkage movements and, if it is to act as a diaphragm and the structure is to act as a unit, some positive connection between wall and roof may be required.

The importance of construction details is well known. Not only do details affect the initial construction cost, but also they have an important influence on the behavior of the building under the traffic of use and the influence of weather. Since the cost of repairs frequently outweighs the cost of construction done properly at the outset, the architect, engineer, and builder should plan all details carefully.

---

*See Ref. 42.

Foundation Details

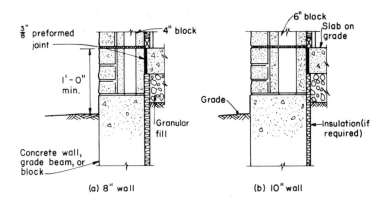

Fig. A-1. Foundation for composite wall.

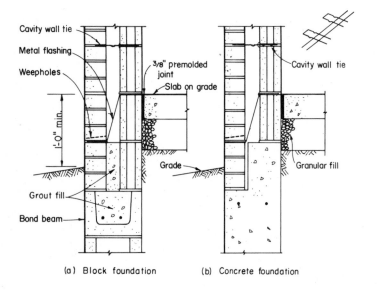

Fig. A-2. Foundation for cavity walls.

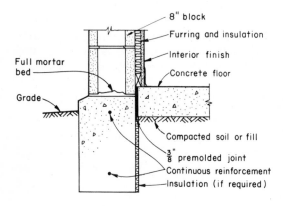

Fig. A-3. Trench-type foundation.

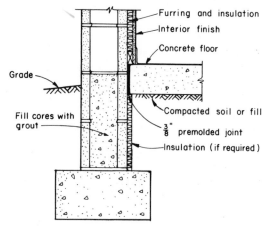

Fig. A-4. Spread footing foundation.

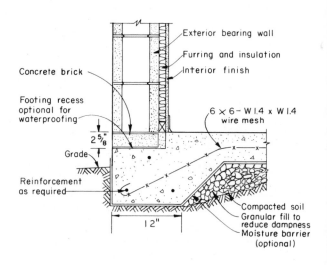

Fig. A-5. Floating slab foundation.

Basement Wall Details

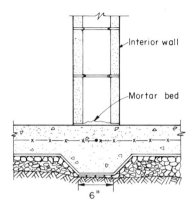

Fig. A-6. Load-bearing wall on slab.

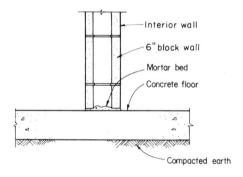

Fig. A-7. Non-load-bearing wall on slab.

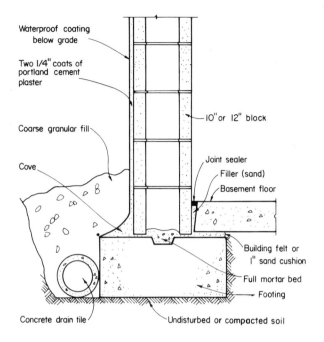

Waterproof coating below grade

Two 1/4" coats of portland cement plaster

Coarse granular fill

Cove

Concrete drain tile

10" or 12" block

Joint sealer

Filler (sand)

Basement floor

Building felt or 1" sand cushion

Full mortar bed

Footing

Undisturbed or compacted soil

Fig. A-8. Exterior wall on footing.

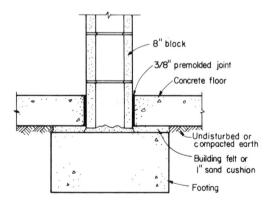

8" block

3/8" premolded joint

Concrete floor

Undisturbed or compacted earth

Building felt or 1" sand cushion

Footing

Fig. A-9. Interior wall on footing.

Wall Details at Precast Concrete Floors

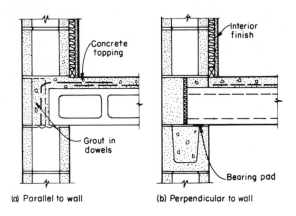

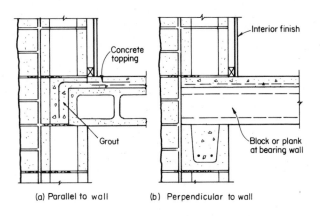

Fig. A-10. Hollow-core slab floors and single-wythe walls.

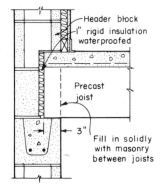

Fig. A-11. Precast joist floor and single-wythe wall.

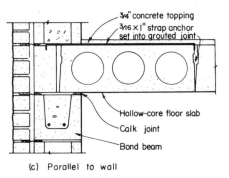

Fig. A-13. Hollow-core slab floor and composite walls.

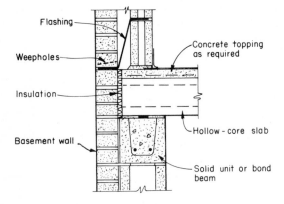

Fig. A-12. Hollow-core slab floors with cavity wall above.

Wall Details at Cast-in-Place Concrete Floors

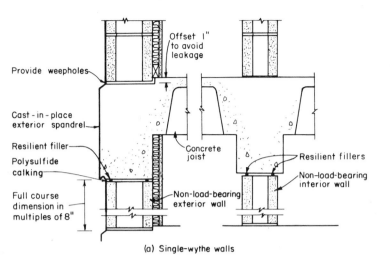

(a) Single-wythe walls

Provide weepholes

Cast-in-place exterior spandrel

Resilient filler

Polysulfide calking

Full course dimension in multiples of 8"

Offset 1" to avoid leakage

Concrete joist

Resilient fillers

Non-load-bearing interior wall

Non-load-bearing exterior wall

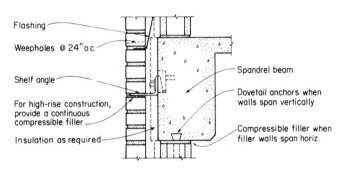

Flashing

Weepholes @ 24" o.c.

Shelf angle

For high-rise construction, provide a continuous compressible filler

Insulation as required

Spandrel beam

Dovetail anchors when walls span vertically

Compressible filler when filler walls span horiz.

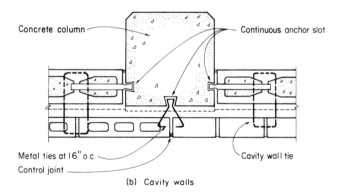

Concrete column

Continuous anchor slot

Metal ties at 16" o.c.

Control joint

Cavity wall tie

(b) Cavity walls

Fig. A-14. Curtain wall and partition with concrete frame.

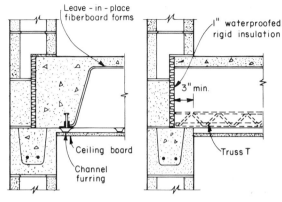

Leave-in-place fiberboard forms

1" waterproofed rigid insulation

3" min.

Ceiling board

Channel furring

Truss T

(a) Single-wythe walls

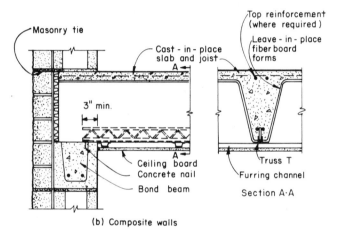

Masonry tie

Cast-in-place slab and joist

Top reinforcement (where required)

Leave-in-place fiber board forms

3" min.

Ceiling board

Concrete nail

Bond beam

Truss T

Furring channel

Section A·A

(b) Composite walls

Fig. A-15. I/D® (Integrated-Distribution) system floors.

®Trademark and servicemark of the Portland Cement Association 1971.

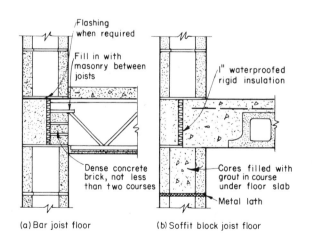

Flashing when required

Fill in with masonry between joists

1" waterproofed rigid insulation

Dense concrete brick, not less than two courses

Cores filled with grout in course under floor slab

Metal lath

(a) Bar joist floor

(b) Soffit block joist floor

Fig. A-16. Joist floors.

Wall Details at Concrete and Metal Roofs

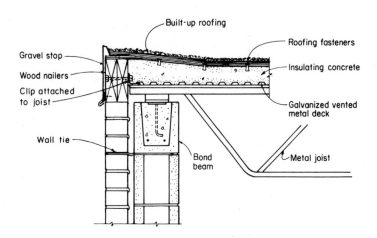

Fig. A-17. Metal deck roof.

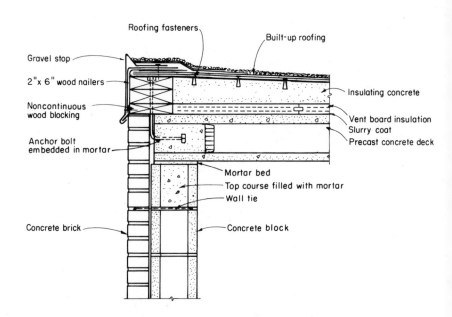

Fig. A-19. Precast concrete deck.

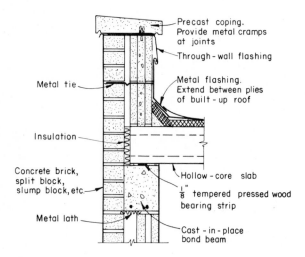

Fig. A-18. Parapet for load-bearing cavity walls.

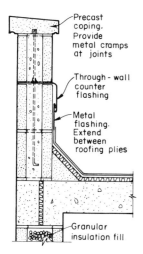

Precast coping. Provide metal cramps at joints

Through-wall counter flashing

Metal flashing. Extend between roofing plies

Granular insulation fill

Fig. A-20. Parapet for flat concrete roof on single-wythe walls.

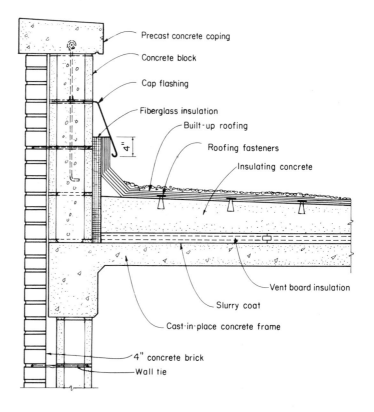

Precast concrete coping

Concrete block

Cap flashing

Fiberglass insulation

Built-up roofing

Roofing fasteners

Insulating concrete

4"

Vent board insulation

Slurry coat

Cast-in-place concrete frame

4" concrete brick

Wall tie

Fig. A-22. Parapet for concrete frame.

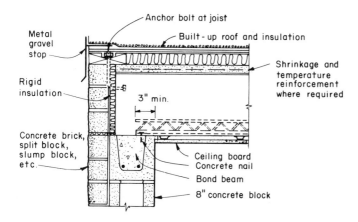

Anchor bolt at joist

Built-up roof and insulation

Metal gravel stop

Shrinkage and temperature reinforcement where required

Rigid insulation

3" min.

Concrete brick, split block, slump block, etc.

Ceiling board
Concrete nail

Bond beam

8" concrete block

Fig. A-21. I/D® roof.

Wall Details at Wood Roofs and Floors

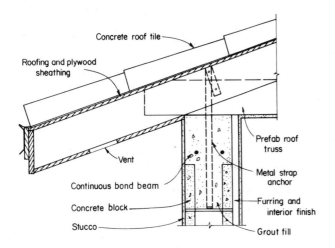

Fig. A-23. Low-sloped wood roofs with bond beam (hurricane-resistant).

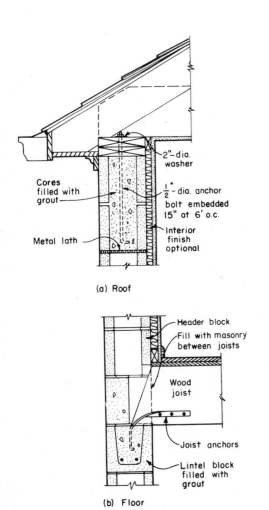

(a) Roof

(b) Floor

Fig. A-24. Steeply sloped wood roof with single-wythe walls.

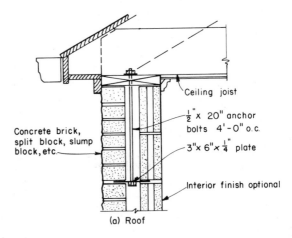

(a) Roof

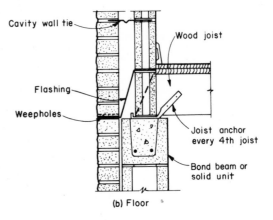

(b) Floor

Fig. A-25. Wood floors and roofs with cavity walls.

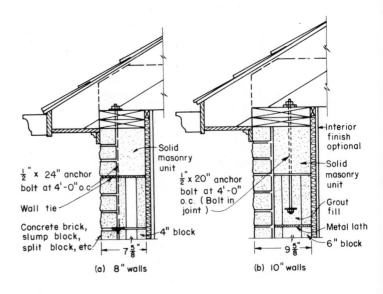

(a) 8" walls   (b) 10" walls

Fig. A-26. Sloped roofs with composite walls.

Door Frame Details

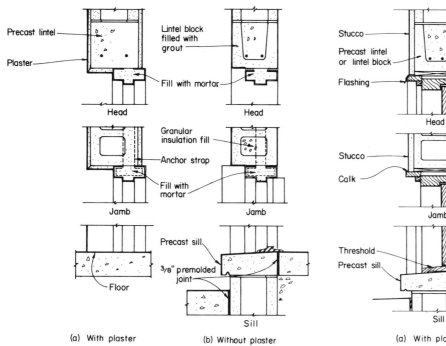

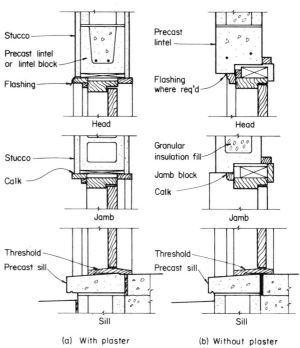

Fig. A-27. Metal door frames with and without plaster.

Fig. A-29. Wood door frames with and without plaster.

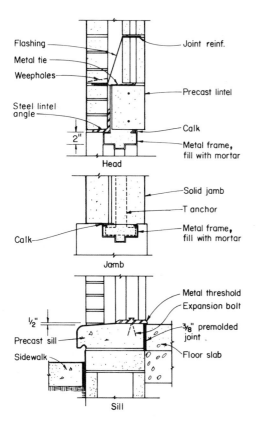

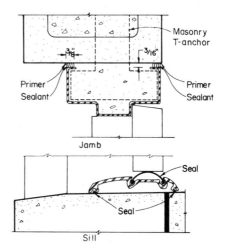

Fig. A-30. Details of metal door frames.

Fig. A-28. Metal door frame with cavity walls.

Metal Window Details

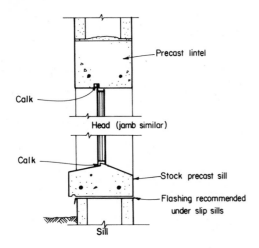

Fig. A-31. Metal basement window.

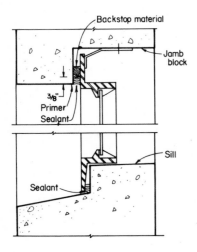

Fig. A-32. Metal window frame.

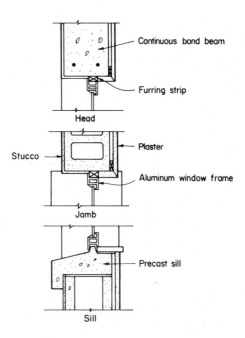

Fig. A-33. Aluminum window with plaster.

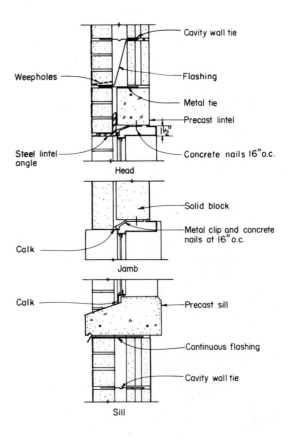

Fig. A-34. Metal windows with cavity walls.

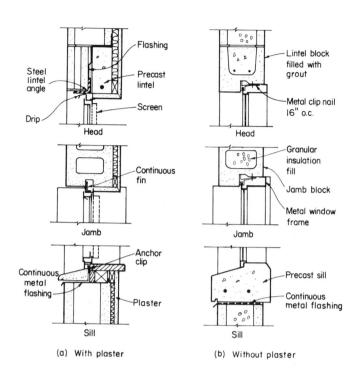

Fig. A-35. Metal windows with and without plaster.

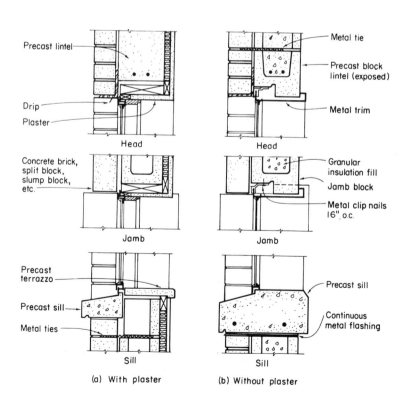

Fig. A-36. Metal windows for composite walls with and without plaster.

Wood Window Details

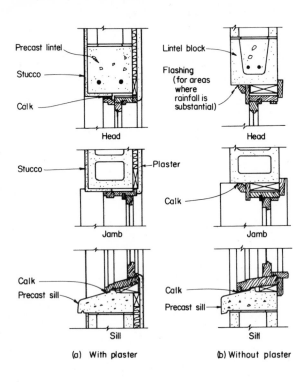

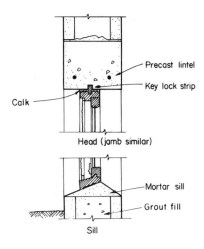

Fig. A-39. Wood basement windows.

Fig. A-37. Double-hung wood windows with and without plaster.

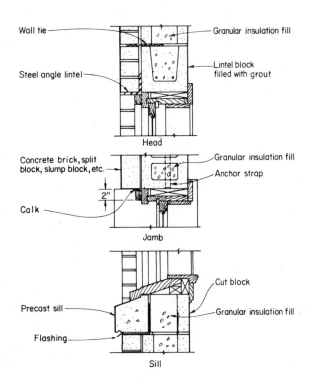

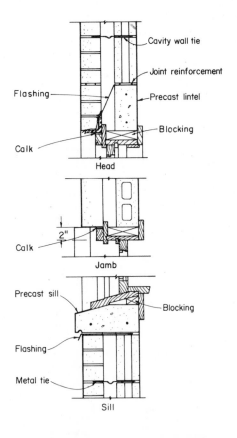

Fig. A-40. Wood windows with cavity wall.

Fig. A-38. Double-hung wood windows with composite wall.

Wall Details for Wood Framing

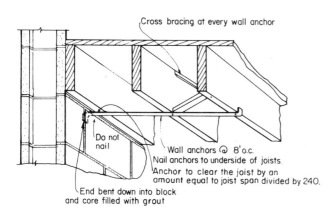

Fig. A-41. Wood joists parallel to wall.

Cross bracing at every wall anchor

Do not nail

Wall anchors @ 8' o.c.
Nail anchors to underside of joists.
Anchor to clear the joist by an
amount equal to joist span divided by 240.

End bent down into block
and core filled with grout

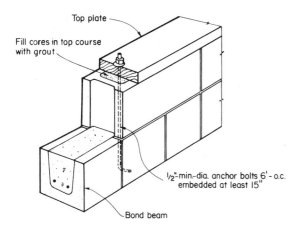

Fig. A-42. Top plate anchorage over wall opening.

Top plate

Fill cores in top course
with grout

1/2"-min.-dia. anchor bolts 6' - o.c.
embedded at least 15"

Bond beam

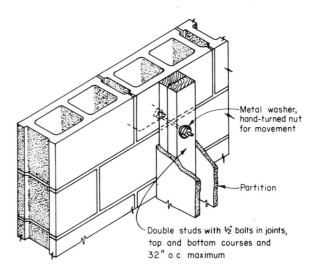

Fig. A-44. Partition anchorage.

Metal washer,
hand-turned nut
for movement

Partition

Double studs with 1/2" bolts in joints,
top and bottom courses and
32" o.c. maximum

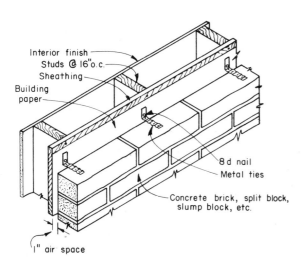

Fig. A-43. Masonry veneer anchorage.

Interior finish
Studs @ 16" o.c.
Sheathing
Building paper

8 d nail
Metal ties

Concrete brick, split block,
slump block, etc.

1" air space

Metric Conversion Factors

The following list will enable readers to convert the U.S. and Canadian customary values of measurements used in this handbook to SI (International System) units, the currently recommended form of the metric system. Also included are a few conversion factors that do not conform strictly to SI but are commonly used in some "metric" nations. The proper conversion procedure is to multiply the specified U.S. or Canadian customary value by the conversion factor exactly as given below and then to round to the appropriate number of significant digits desired. For example, to convert 11.4 ft. to meters: 11.4 X 0.3048 = 3.47472, which rounds to 3.47 meters for an accuracy of two significant digits. Do not round either value before performing the multiplication, as accuracy would be reduced. A complete guide to the SI system and its use can be found in ASTM E380, Standard Metric Practice Guide (A Guide to the Use of SI—the International System of Units).

To convert from	to	multiply by
Length		
inch (in.)	centimeter (cm.)	2.54 E*
inch (in.)	meter (m.)	0.0254 E
foot (ft.)	meter (m.)	0.3048 E
yard (yd.)	meter (m.)	0.9144 E
Area		
square foot (sq.ft.)	square meter (sq.m.)	0.09290
square inch (sq.in.)	square centimeter (sq.cm.)	6.452
square inch (sq.in.)	square meter (sq.m.)	0.0006452
square yard (sq.yd.)	square meter (sq.m.)	0.8361
Volume		
cubic inch (cu.in.)	cubic centimeter (cu.cm.)	16.39
cubic inch (cu.in.)	cubic meter (cu.m.)	0.00001639
cubic foot (cu.ft.)	cubic meter (cu.m.)	0.02832
cubic yard (cu.yd.)	cubic meter (cu.m.)	0.7646
gallon (gal.) Can. liquid**	liter	4.546
gallon (gal.) Can. liquid**	cubic meter (cu.m.)	0.004546
gallon (gal.) U.S. liquid**	liter	3.785
gallon (gal.) U.S. liquid**	cubic meter (cu.m.)	0.003785
Force		
kip	kilogram (kgf)	453.6
kip	newton (N)	4,448.
pound (lb.)	kilogram (kgf)	0.4536
pound (lb.)	newton (N)	4.448
Pressure or Stress		
kip per square inch (ksi)	kilogram per square centimeter (kg/sq.cm.)	70.31
pound per square foot (psf)	kilogram per square meter (kg/sq.m.)	4.882
pound (force) per square foot (psf)	pascal (Pa.)†	47.88
pound per square inch (psi)	kilogram per square centimeter (kg/sq.cm.)	0.07031
pound (force) per square inch (psi)	pascal (Pa.)†	6,895.

The prefixes and symbols listed are commonly used to form names and symbols of the decimal multiples and submultiples of the SI units.

Multiplication Factor	Prefix	Symbol
$1\,000\,000\,000 = 10^9$	giga	G
$1\,000\,000 = 10^6$	mega	M
$1\,000 = 10^3$	kilo	k
$1 = 1$	—	—
$0.001 = 10^{-3}$	milli	m
$0.000\,001 = 10^{-6}$	micro	μ
$0.000\,000\,001 = 10^{-9}$	nano	n

To convert from	to	multiply by
Mass (Weight)		
pound (lb.) avdp.	kilogram (kg)	0.4536
ton, 2,000 lb.	kilogram (kg)	907.2
grain	kilogram (kg)	0.00006480
Mass (Weight) per Length		
kip per linear foot (klf)	kilogram per meter (kg/m.)	0.001488
pound per linear foot (plf)	kilogram per meter (kg/m.)	1.488
Mass per Volume (Density)		
pound per cubic foot (pcf)	kilogram per cubic meter (kg/cu.m.)	16.02
pound per cubic yard (pcy)	kilogram per cubic meter (kg/cu.m.)	0.5933
Temperature		
degree Fahrenheit (deg. F.)	degree Celsius (C)	$t_C = (t_F - 32)/1.8$
degree Fahrenheit (deg. F.)	degree kelvin (K)	$t_K = (t_F + 459.7)/1.8$
Energy		
British thermal unit (Btu)	joule (J)	1,056.
kilowatt-hour (kwh)	joule (J)	3,600,000. E
Power		
horsepower (hp) 550 ft.-lb./sec.	watt (W)	745.7
Velocity		
mile per hour (mph)	kilometer per hour	1.609
mile per hour (mph)	meter per second (m./s.)	0.4470

*E indicates that the factor given is exact.
**One U.S. gallon equals 0.8327 Canadian gallon.
†A pascal equals 1.000 newton per square meter.

Related Materials

The following PCA materials may be of interest to readers of the
Concrete Masonry Handbook.

Publications

EB049M	Portland Cement Plaster (Stucco) Manual
EB086B	Building Movements and Joints
IS040M	Mortars for Masonry Walls
IS159T	Acoustics of Concrete in Buildings
IS181M	Masonry Cement Mortars
IS219M	Permeability Tests of Masonry Walls
IS220M	Building Weather-Resistant Masonry Walls
PA043M	Recommended Practices for Laying Concrete Block
RD066M	Sound Transmission Loss Through Concrete and Concrete Masonry Walls
RD067M	Behavior of Inorganic Materials in Fire
RD071M	Thermal Performance of Masonry Walls
RD075M	Heat Transfer Characteristics of Walls Under Dynamic Temperature Conditions
SR205B	Luxury Apartment Developer Depends on Concrete Masonry and Prestressed Concrete
SR218B	Concrete Products Give Investment Edge to Apartment Developer
SR224B	Passive Solar House Designs Meet Consumer Needs
SR229B	Concrete Products for Two-Story Apartment Buildings

Motion Picture

PC010	ABC's of Concrete Masonry Construction

Slide Sets

CS001	Concrete Brick Veneers
CS007	Concrete Masonry Basements for Better Homes
CS008	Building with Concrete Brick Veneers
SS017	Masonry Load-Bearing Walls
SS018	Quality Concrete Masonry Construction Details
SS021	Control Joints for Concrete Masonry
SS081	Acoustics of Concrete in Buildings

To order, write or call Order Processing, Portland
Cement Association, 5420 Old Orchard Road,
Skokie, Illinois 60077-4321. Phone 312/966-6200,
ext. 450.

Descriptions of the materials mentioned above as
well as a complete listing of PCA publications,
motion pictures, and slide sets can be found in the
Catalog of Publications and Audiovisual Materials
(MS254G). The catalog is free and available upon
request.